# Current Topics in Microbiology and Immunology

Volume 433

The reviews series *Current Topics in Microbiology and Immunology* publishes cutting-edge syntheses of the latest advances in molecular immunology, medical microbiology, virology and biotechnology. Each volume of the series highlights a selected timely topic, is curated by a dedicated expert in the respective field, and contains a wealth of information on the featured subject by combining fundamental knowledge with latest research results in a unique manner.

2019 Impact Factor: 3.095., 5-Year Impact Factor: 3.895

2019 Eigenfaktor Score: 0.00081, Article Influence Score: 1.363

2019 Cite Score: 6.0, SNIP: 1.023, h5-Index: 43

More information about this series at http://www.springer.com/series/82

Harvinder Singh Gill · Richard W. Compans
Editors

# Nanoparticles for Rational Vaccine Design

Responsible Series Editor: Richard W. Compans

Springer

*Editors*
Harvinder Singh Gill
Department of Chemical Engineering
Texas Tech University
Lubbock, TX, USA

Richard W. Compans
Department of Microbiology
and Immunology
Emory University
Atlanta, GA, USA

ISSN 0070-217X ISSN 2196-9965 (electronic)
Current Topics in Microbiology and Immunology
ISBN 978-3-030-85066-1 ISBN 978-3-030-85067-8 (eBook)
https://doi.org/10.1007/978-3-030-85067-8



This Springer imprint is published by the registered company Springer Nature Switzerland AG
The registered company address is: Gewerbestrasse 11, 6330 Cham, Switzerland

# Preface

With emerging infectious diseases and increased zoonotic transmissions, strategies for rapid vaccine development are required to prevent disease outbreaks. Based on genetic analysis of pathogens, their surface antigens can be identified; however, the subsequent steps of creating a protective vaccine requires careful consideration. Scientific advancements in divergent fields including nanotechnology, immunology, engineering, and biotechnology have spawned platform technologies that can be rapidly deployed to create vaccine candidates. A unifying theme in all these platforms is that they comprise tiny particles, called nanoparticles.

Nanoparticles are often a few hundred nanometers in size or smaller, and can stimulate the immune system in unique ways to generate robust humoral and cellular immune responses. The basis of this robust stimulation by nanoparticles is multifold, including the ability of nanoparticles to engage immune cells in a manner that enhances antigen presentation, secretion of proinflammatory signals, and engagement of secondary lymphoid tissues by directly transporting particles to these sites through the lymphatics.

In this volume of *Current Topics in Microbiology and Immunology*, four major nanoparticle technologies are presented. The Chap. 1 begins by discussing one of the earliest technologies which was based on liposomes. This chapter reviews the history and fate of liposomes via the major histocompatibility processing pathways, and summarizes some of the human liposomal vaccine trials. The Chap. 2 describes polymeric particle technology and considers the adjuvant effect of nanoparticles, with discussion of effects of nanoparticle size, shape, surface charge and hydrophobicity on the immune response. This chapter summarizes both natural and human-synthesized polymers used for vaccine development and describes their uses in vaccine development against pathogens and cancer. The Chap. 3 describes the use of virus-like particles as a vaccination platform. The virus-like particle platform was first used to make the hepatitis B vaccine and most recently the human papillomavirus vaccine. The chapter discusses the production system of virus-like particles and highlights their use for vaccine development against many disease-causing agents including influenza virus, respiratory syncytial virus, *Toxoplasma gondii*, and malaria. Emphasis is placed on discussion of preclinical as well as clinical studies. In the last chapter, a more recent approach of creating nanoparticles purely out of subunit antigens is discussed. In this approach no carrier substances are used, and instead the antigenic protein is converted into a nanoparticle called a protein nanocluster. This prevents off target immune responses against carriers and focuses the immune response towards the target antigens. The chapter discusses strategies for making protein nanoclusters, their trafficking and uptake by antigen presenting cells, and humoral and cellular responses generated by their use in vivo.

This volume is intended as a teaching aid to provide a review of the current nanoparticle technologies that have seen significant growth over the past decades and are uniquely positioned to make an impact.

Lubbock, TX, USA
Atlanta, GA, USA
July 2021

Harvinder Singh Gill
Richard W. Compans

# Contents

# Liposome Formulations as Adjuvants for Vaccines

**Mangala Rao, Kristina K. Peachman, and Carl R. Alving**

## Contents



**Abstract** Development of liposome-based formulations as vaccine adjuvants has been intimately associated with, and dependent on, and informed by, a fundamental understanding of biochemical and biophysical properties of liposomes themselves. The Walter Reed Army Institute of Research (WRAIR) has a fifty-year history of experience of basic research on liposomes; and development of liposomes as drug

---



---

M. Rao (✉)
Chief, Laboratory of Adjuvant & Antigen Research, U.S. Military HIV Research Program, Walter Reed Army Institute of Research, 503 Robert Grant Avenue, Silver Spring, MD 20910, USA
e-mail: mrao@hivresearch.org; mangala.rao.civ@mail.mil

K. K. Peachman · C. R. Alving
Laboratory of Adjuvant & Antigen Research, U.S. Military HIV Research Program, Walter Reed Army Institute of Research, 503 Robert Grant Avenue, Silver Spring, MD 20910, USA

Current Topics in Microbiology and Immunology (2021) 433: 1–28
https://doi.org/10.1007/82_2020_227

Published Online: 10 November 2020

carriers; and development of liposomes as adjuvant formulations for vaccines. Uptake of liposomes by phagocytic cells in vitro has served as an excellent model for studying the intracellular trafficking patterns of liposomal antigen. Differential fluorescent labeling of proteins and liposomal lipids, together with the use of inhibitors, has enabled the visualization of physical locations of antigens, peptides, and lipids to elucidate mechanisms underlying the MHC class I and class II pathways in phagocytic APCs. Army Liposome Formulation (ALF) family of vaccine adjuvants, which have been developed and improved since 1986, and which range from nanosize to microsize, are currently being employed in phase 1 studies with different types of candidate vaccines.

## 1 Introduction

The concept of using vesicles constructed from purified lipids, now known as liposomes, as models of cell membranes was introduced by (Bangham et al. 1965). Although many types of lipids can form liposomes, because of the similarity of liposomal phospholipid membranes to those of cells, the term "liposome" originally referred to vesicles which contained bulk lipids comprising phospholipids, with or without cholesterol. The discovery of liposomes led to three major directions of subsequent research: basic research on membranes containing lipid bilayers (Pick et al. 2018); development of liposome-encapsulated drugs for enhanced treatment of enzyme deficiencies, cancer, diabetes, or other diseases (Gregoriadis 1978; Li et al. 2019); and development of liposome-based vaccines (Schwendener 2014; De Serrano and Burkhart 2017; Wang et al. 2019; Alving et al. 2020). Thus, liposomes have been used both for prophylactic purposes as carriers of vaccine adjuvants and for therapeutic purposes as drug carriers (Nisini et al. 2018).

For use of liposomes as a vaccine formulation, the net charge of the particle should be considered, and the liposomes are generally either negative (anionic liposomes) or positive (cationic liposomes). Cationic liposomes are not addressed in this review but are reviewed in (Pedersen et al. 2018; De Serrano and Burkhart 2017). Early in the liposome field, two important discoveries dominated the direction of translational research: First, upon exposure to human serum, liposomes bound huge amounts of serum proteins, especially complement proteins; and second, upon intravenous injection into experimental animals, liposomes were taken up almost exclusively by macrophages in the liver and spleen.

The first application of liposomes as a vaccine adjuvant formulation in humans utilized a recombinant malaria protein which was encapsulated in multilamellar liposomes containing saturated neutral and anionic phospholipids, cholesterol, and monophosphoryl lipid A (MPLA) as an adjuvant (Fries et al. 1992). Although it was believed at that time that the liposomal protein was fully encapsulated, it is now believed that some of the proteins were also located both as transmembrane protein and as surface-adsorbed protein (Verma et al. 1991). Because of this, in this review we refer to proteins either in or on liposomes only as "liposome-associated" proteins. These liposomes represented the first iteration of an adjuvant composition that is now further refined and known as Army Liposome Formulation (ALF) (Beck et al. 2015b). The above liposomes containing malaria antigen were adsorbed to aluminum salt as a second adjuvant, and the formulation was injected intramuscularly (Fries et al. 1992). This approach, utilizing a mixture containing a range of large (microsize) and small (nanosize) liposomal particles, together with two different adjuvants for stimulation of innate immunity, seemed suitable for achieving vaccine potency in humans, and in three phase 1 clinical trials it did lead to high titers of antibodies to two different recombinant malaria protein antigens (Fries et al. 1992; Heppner et al. 1996) and to a recombinant HIV envelope protein, gp120 (Rao et al. 2018). These early studies utilizing liposomes as carriers of vaccines along with adjuvants laid the groundwork for future development of licensed liposomal vaccines.

A commercial GlaxoSmithKline (GSK) liposome formulation known as AS01B, which contains bulk lipids consisting of unsaturated neutral phospholipids and cholesterol, and which has both MPLA and a triterpenoid saponin known as QS21 as immunostimulants, has been included in an EMA-approved malaria vaccine in Africa (Mosquirix®, GSK) and in an FDA-licensed shingles vaccine (Shingrix®, GSK) in the US (Didierlaurent et al. 2017; Alving et al. 2020). The evolution of liposomes containing phospholipids, cholesterol, and MPLA and other immunostimulants as adjuvants thus presents a useful model for rational development of liposomes as carriers of antigens and adjuvants for vaccines. Various physical factors of liposomes, such as: surface charge (neutral, anionic, cationic); size (nano vs. micro); phospholipid fatty acyl composition (saturated vs. unsaturated); molar ratios of cholesterol and phospholipid in the lipid bilayer; number of lipid bilayer lamella (unilamellar, oligolamellar, or multilamellar vesicles); targeting molecules for attachment to immune cells; and the type, composition and number, if any, of attached adjuvant(s); have all been identified as factors related to the ease and cost of manufacture, and potential for unwanted toxicity. Putting all of these variables together has generated a complex number of ingenious physical immunostimulant compositions (Wang et al. 2019). Nonetheless, regardless of the structures of different types of liposomes, they are all foreign materials in the body, and each of the different liposome formulations faces the challenge of activation of innate immunity which has as its main goal the removal of the foreign particle.

## 2 Binding of Complement and Other Proteins to Liposomes Leading to Opsonization

In 1968, Kinsky and colleagues demonstrated that antigen-specific antibodies could bind to a liposomal glycolipid membrane antigen, resulting in complement-dependent damage and release of encapsulated liposomal glucose (Haxby et al. 1968, 1969; Kinsky et al. 1969). Further work demonstrated that the concentrations, by weight, of liposome-bound proteins were huge (as much as 800 μg of bound protein per μmole of liposomal phospholipid), and the amount of protein was easily measured by a Lowry protein assay on the surface of washed liposomes (Alving and Kinsky 1971). The above studies initiated the field of immunological aspects of liposomes; and they also introduced the concept that serum proteins (such as antibodies and complement) could stick to liposomes and accumulate in large amounts on the membrane surface.

In the early days of research on immunological aspects of liposomes, complement was thought to consist mainly of nine serum components (Nelson et al. 1966), and "complement" was, and often is today, defined informally as "fresh, unheated serum." Incubation of liposomes containing an antigen, which were then coated with specific antibodies bound to the antigen, and then exposed to complement (fresh unheated serum), resulted in such a spectacular degree of phagocytic uptake in vitro by mouse peritoneal macrophages that the cells became rounded and bulging with slowly degrading intracellular vesicles (Wassef and Alving 1987).

In 1986, it was shown that addition of suspensions of liposomes lacking an antigen to human serum resulted in a more complex protein coating that included the nonspecific binding of immunoglobulins, complement proteins, albumin, fibrinogen, fibronectin, C-reactive protein, certain serum clotting factors, and scores of other proteins (Bonte and Juliano 1986; Chonn et al. 1991, 1992). In addition, it was also demonstrated that normal human serum contains naturally occurring antibodies to numerous phospholipids and cholesterol, and that the natural antibodies can bind to liposomes containing those components (Alving 1984; Alving and Wassef 1999; Matyas and Alving 2011). Upon intravenous injection of liposomes for the purpose of delivering drugs, the proteins that were adsorbed to the surface of liposomes in vivo resulted in the uptake of the injected particles by phagocytic cells, mainly Kupffer cells in the liver and splenic macrophages (Segal et al. 1974; Tyrrell et al. 1976). From those studies, it became apparent that incubation of liposomes in fresh human serum results in opsonization of the liposomes with numerous types of proteins, particularly the complement proteins, resulting in uptake of the liposomes by phagocytic cells.

In 1974, it was demonstrated that liposomes could serve as carriers of protein antigens, or even haptens, for immunization (Allison and Gregoriadis 1974; Uemura et al. 1974). The affinity of liposomes for uptake by phagocytic cells as antigen presenting cells has served as a major justification and rationale for using liposomes as vehicles for vaccines (Alving 1991). For liposome formulations as vaccine constituents, delivery of liposomal antigen to macrophages, dendritic cells, and T and B lymphocytes may be required for effectiveness.

It is now known from research on nanoparticles which are used for pharmaceutical and medical applications that exposure of liposomes and other types of particles to biological fluids in vivo or in vitro results in adsorption of proteins that alter the chemical and physical characteristics of the particles, a surface coating phenomenon that is referred to in the nanomedicines field as a "protein corona" (Cedervall et al. 2007; Ke et al. 2017). As noted by Ke et al. (2017): "… the protein corona owes its presence to thermodynamics in an aqueous environment, especially to the minimization of free enthalpy, and is mediated by Coulombic and van der

Waals forces, hydrogen bonding, and hydrophobic interactions. Upon introduction into a biological fluid, a nanoparticle first assumes a transient or "soft" corona rendered by proteins of high abundance and is subsequently coated over time by a "hard" corona or proteins of high affinity according to the Vroman effect."

From these new insights, it is useful to consider the influence of surface-bound protein on liposomal carriers of antigens and adjuvants for vaccines. Although numerous proteins can attach to liposomes to form a protein corona, many with variable affinities, it seems reasonable from the earliest liposome research, and from recent research on nanoparticles (Vu et al. 2019), to suspect that IgG and IgM immunoglobulins, whether natural antibodies to the lipids or to other attached proteins in the corona, would activate the complement cascade. As described by Bohlson et al.: "The complement system is composed of more than 50 different molecules and cleavage products, including, but not limited to, pattern recognition molecules, proenzymes, proteases, anaphylatoxins, opsonins, receptors, regulators, and multi-molecular complexes that are critical to host defense and maintenance of normal tissue homeostasis." (Bohlson et al. 2019). Three pathways to complement activation have been identified, each initiated by different recognition molecules and mechanisms: (A) *classical pathway*, mainly initiated by antibody binding to an antigen, or binding other molecules, leading to recognition by C1q; (B) *lectin pathway*, initiated by recognition of ligands by mannose-binding lectin (MBL) and several other molecules; and (C) the *alternative pathway*, which can be activated spontaneously, although properdin may play a role (Lubbers et al. 2017). In an evolutionary sense, complement is thought to have evolved as a mechanism of innate immunity that can be activated by appearance of foreign invaders to produce protective effector mechanisms. It is likely that the natural antibodies to autologous lipids, proteins, or nucleic acids exist as markers to recognize cellular debris, or for recognizing the appearance of epitopes that are normally hidden due to steric hindrance; and the antibodies serve as agonists that attract scavenger cells (Alving and Wassef 1999).

The protein corona can pose a difficult problem for targeted intravascular delivery of liposomal therapeutic drugs, for example, to tumor cells, because the liposomes are rapidly coated with opsonins, including complement, resulting in uptake by phagocytic cells. However, from a vaccine standpoint, delivery of liposomes that are coated with opsonins (such as complement) to cells in the immune system, including phagocytic cells, represents a natural advantage and also highlights a rational mechanism for using liposomes for adjuvant development.

The composition of the protein corona that will associate with the liposomes after injection in vivo depends on several factors. First, the initial protein corona, which is the only protein or peptide initially associated with the liposomes, is the antigen itself. The affinity of the immunogenic liposome-associated peptide or protein for the liposomal surface can be influenced by the liposomal surface charge, by the hydrophobicity of the antigen and by its access to the hydrophobic region of the lipid bilayer, or by the composition of the liposomal phospholipid headgroup.

# 3 Intramuscular Fate of Injected Liposomal Adjuvant Formulations

Among the many mechanisms by which liposomal vaccine adjuvants work, targeting of the liposomes to phagocytic cells, and activation of the cells can be an important and sometimes a critical first step for initiation of the adaptive immune response. The original paradigm in the complement field held that complement was a system of proteins that were present in serum. However, certain locations in the body exist where serum proteins do not have ready access, but where complement or other types of opsonins are still needed, for example, in skeletal muscle or subcutaneous tissue, or in other areas that include interstitial spaces. By the same token, skeletal muscle contains very few immune cells such as macrophages, monocytes, dendritic cells, and neutrophils (Liang et al. 2015). Recent research has demonstrated that various complement components, regulators, and receptors are synthesized by immune cells themselves, including polymorphonuclear leukocytes, mast cells, monocytes, macrophages, dendritic cells, NK cells, and B and T lymphocytes (Lubbers et al. 2017). In addition, a "hidden arsenal" of complement and complement-related proteins exists inside T lymphocytes, and perhaps in many, and perhaps even most cells (Liszewski et al. 2017). However, the major problem for vaccine development is that immune cells are only few in number in muscle tissue. For liposomes that bind directly to cells, such as certain types of targeted liposomal (virosome) vaccines (Zurbriggen 2003; Herzog et al. 2009), or for cationic liposomes that attach to negatively charged cell surfaces to form a depot (Henriksen-Lacey et al. 2010), this might not be a great problem.

It is interesting to speculate that perhaps complement (or natural antibodies), which are synthesized by even a few cells in skeletal muscle, might accumulate in the local extracellular environment; and these materials might bind to injected liposomes, resulting in complement activation and generation of chemotactic factors and other attractants that could result in increased numbers of immune cells through chemotaxis. However, because of the paucity of immune cells in skeletal muscle, and the paucity of opsonins in muscle that attach to liposomes, in many instances anionic liposomes by themselves lacking adjuvants do not increase the immunogenicities of certain antigens, and the liposomal formulation requires addition of an adjuvant, such as monophosphoryl lipid A (Alving et al. 1986; White et al. 1995). Thus, it is more likely that an immunostimulator that serves as an adjuvant must be added to many types of vaccines, including neutral or anionic liposomal vaccines, to cause a local inflammation in order to achieve a significant adaptive immune response to the antigen after injection in skeletal muscle (Liang and Lore 2016). Even second-generation virosomal virus-like particle types of liposomal vaccines are greatly improved by the addition of an adjuvant (Kamphuis et al. 2012; Moser et al. 2013; Smith et al. 2013). Regardless of the mechanism by which liposomal formulations containing adjuvants attract immune cells to the intramuscular injection site, the intracellular fate of liposomes in phagocytic cells is often a critical step in the adaptive immune response.

## 4 Intracellular Fate of Liposomes in Phagocytic Cells

As mentioned above, there is an influx of phagocytic cells to the intramuscular injection site and, after uptake of the liposome-associated antigen by a phagocyte the antigen is ultimately transported to the draining lymph nodes for interaction with T-cells. One of the main roles of phagocytic antigen presenting cells (APC) is to sample the environment and present antigens, and then to start appropriate immune signaling once the danger is identified. The first step in this process is the internalization of the liposomal antigen by phagocytic cell for processing followed by presentation of the peptides on the surface of the antigen presenting cell in the context of the major histocompatibility complex (MHC) class I or class II molecules to interact with the T-cell receptor. Antigen processing is highly complex, involving a number of different components, some of which overlap between MHC class I and class II pathways. In conjunction with the secondary stimulus consisting of costimulatory molecules, an antigen-specific cellular (Th1) or a humoral (Th2) response is induced along with the corresponding cytokines.

Liposomes, both anionic and cationic, induce mixed Th1/Th2 responses (Rao et al. 1999a, 2002; Jafari et al. 2018). Antigen processing through both the MHC class I and class II pathways can account for the mixed Th1/Th2 responses induced by liposomes containing MPLA-associated antigen. Typically, endogenous antigens are processed through the MHC class I pathway, while exogenous antigens are processed through the MHC class II pathway (Bevan 1987; Germain and Margulies 1993). Liposomes containing antigen and MPLA are unique in that, although liposomes are exogenous particles, liposomal antigen can escape to the cytoplasm, as shown by the presence of gold-labeled antibody–antigen complex in the cytoplasm (Fig. 1). The antigen can be processed through both the MHC class I (Fig. 2) and class II pathways (Fig. 3). The antigen presentation pathway leading to the loading of exogenous antigens on MHC class I molecules is called cross-presentation. Cross-presentation can occur by two main pathways; the cytosolic pathway, in which antigen processing occurs in the cytosol, and the vacuolar pathway, in which antigen processing occurs within endocytic compartments (Joffre et al. 2012). Because of the close contact between the phagosomes and the endoplasmic reticulum (ER), it has been speculated that specific lipids might act as second messengers promoting a cross-presentation pathway, although no such lipids have been identified so far (Nunes-Hasler and Demaurex 2017). In contrast to macrophages, the endocytic compartments of DCs prevent destruction of internalized antigens due to limited proteolysis; and because of this, DCs, but not macrophages, favor cross-presentation.

At one time, it was proposed that "pH-sensitive" liposomes were a requirement for the delivery of exogenous antigen into the cytoplasmic compartment of the APCs and for the induction of cytotoxic T-cells (CTLs) in vivo (Reddy et al. 1991; Collins et al. 1992). However, it is now well established that liposomes which are pH-insensitive can also deliver the antigen to the cytosol (Fig. 1) and can induce good CTL responses against a variety of antigens, ranging from viruses to

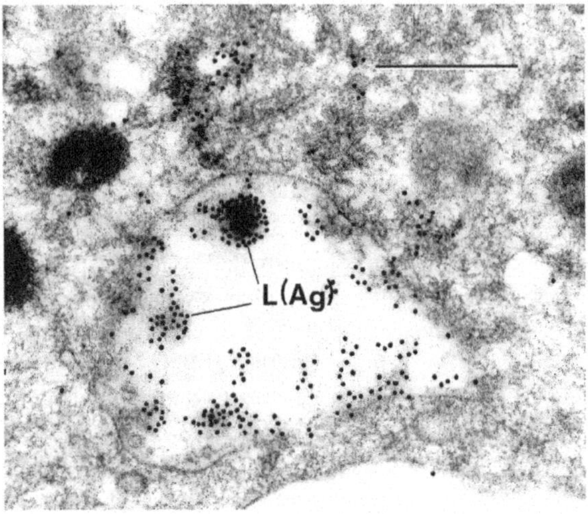


**Fig. 1** Immunogold electron microscopy of macrophages after phagocytosis of liposomes containing lipid A (isolated from *Salmonella minnesota* R595 LPS) and recombinant R32NS1 malaria antigen. Murine bone marrow-derived macrophages were fixed 6 h after incubation with liposomes containing malaria antigen. The malaria antigen was detected by a specific antibody to the malaria antigen followed by treatment with gold-labeled secondary antibody. Please note that the gold-labeled antibody–antigen complex has escaped into the cytoplasm. Bar = 0.5 μm. From Verma et al. (1991)

encapsulated or surface-bound recombinant proteins and peptides (Engler et al. 2004; Rao et al. 2002, 2004; White et al. 1993, 1995; Laborde et al. 2017). To demonstrate the presence of liposomal antigens in the various cellular compartments, we utilized bone marrow-derived macrophages as pure populations of APCs and studied the intracellular trafficking of liposome-encapsulated antigen in vitro, using fluorescently labeled lipids and proteins (Rao and Alving 2000).

## *4.1 MHC Class I Processing Pathway*

Multiple steps within the MHC class I and II pathways have been characterized for liposome-associated antigen uptake and processing. The first step in each pathway is the internalization of the liposome-associated antigen either through pinocytosis, endocytosis, or phagocytosis, depending upon the size of the liposomes. For further details on a variety of internalization mechanisms of nanostructured systems, see (Voltan et al. 2017). These different routes of uptake could result in the antigen entering different processing compartments such as the endosomes, lysosomes, or the cytosol, as shown pictorially in Figs. 2 and 3 (steps 1 and 2). The presence of liposomal antigen in the cytoplasmic compartment has been shown for ovalbumin, ferritin or a malaria antigen by electron microscopy or by the use of fluorescent

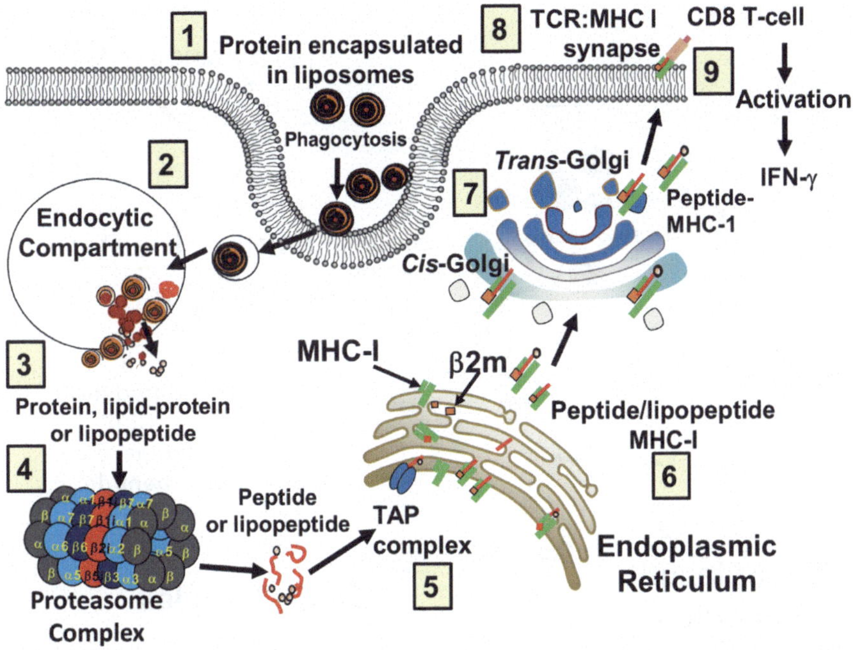


**Fig. 2** MHC class I liposomal antigen trafficking pathway. (1) Protein encapsulated in liposomes is phagocytosed into antigen presenting cells; and (2) enters an endocytic compartment. (3) Protein, lipid-protein or lipopeptide are released into the cytosol of the cell, and (4) are processed by either a proteasome or immunoproteasome complex. (5) Peptide or lipopeptide is then transported by the heterodimeric TAP complex into the endoplasmic reticulum and loaded onto MHC class I-β2 microglobulin molecules. (6) The MHC class I-loaded molecule traffics to the cis-Golgi. (7) From the cis-Golgi, the complex traffics through trans-Golgi complex. (8) After exiting the trans-Golgi, the MHC class I-peptide complex is transported to the cell surface where it binds to $CD8^+$ T-cells via the T-cell receptor. (9) The TCR-MHC-I/peptide complex activates the $CD8^+$ cell, which in turn releases cytokines such as IFN-γ

liposomes (Zhou et al. 1994; Perry and Martin 1995; Verma et al. 1991; Rao and Alving 2000). At the point of contact between either a soluble or liposomal antigen with the cell membrane, a rearrangement of the cytoskeleton occurs to facilitate the uptake of the antigens. Further trafficking of the liposomal antigen requires a functional microtubule-dependent translocation system (Peachman et al. 2004). The protein, lipoprotein or lipopeptide that escapes into the cytoplasm (Fig. 2 step 3) is ubiquitinated and then channeled through the proteasome complex (Fig. 2, step 4) for further hydrolyses into short peptides (Rothwell et al. 2000; Steers et al. 2008).

Peptides that are generated by the proteasome cleavage step are transported across the membrane of the ER and the cis- and trans-Golgi in an ATP-dependent manner by specific trimeric transporters associated with antigen processing (TAP) (Fig. 2 steps 5–7 and Fig. 4A). TAP proteins consist of TAP1 and TAP2,

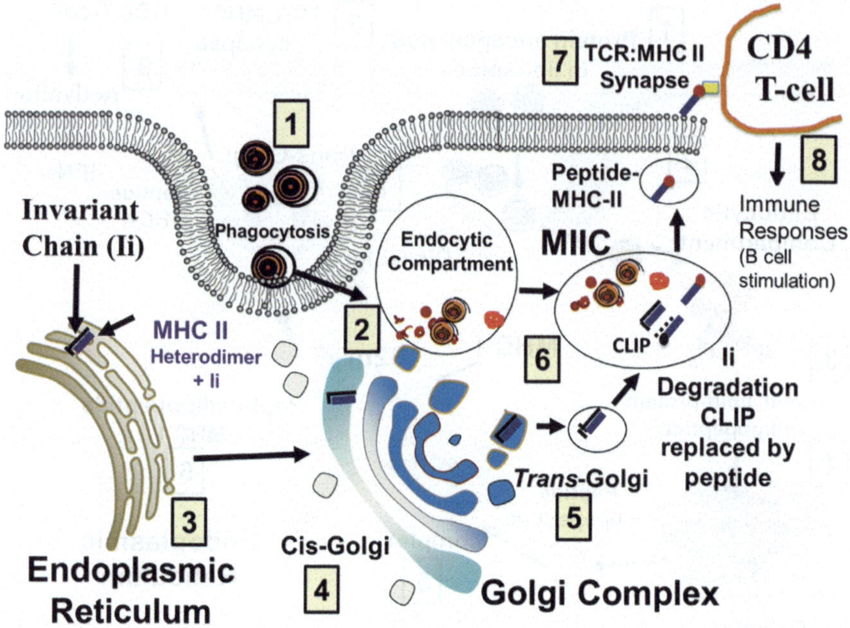


**Fig. 3** MHC class II liposomal antigen trafficking pathway. (1) Liposome-associated antigen is phagocytosed into antigen presenting cells, and (2) enters an endocytic compartment. (3) The heterotrimeric α and β chains of the MHC class II molecule along with the invariant chain (Ii) are synthesized in the endoplasmic reticulum, and traffic to the (4) cis- and (5) trans-Golgi complex. (6) Following maturation in the Golgi complex, the MHC class II heterotrimeric complex with the Ii chain bound to the MHC class II groove then enters the MHC class II compartment (MIIC) where the invariant chain is degraded into the Class II Invariant chain Peptide (CLIP) which is still bound to the MHC-II groove. Peptides generated in the endosomes from exogenous liposomal antigens also enter the MIIC compartment. In this compartment, CLIP (present in the MHC class II groove) is replaced by the peptides (generated in the endosomes) that also traffic to the MIIC compartment. (6) The peptide-MHC-II complexes then move to the cell surface, (7) for interaction with the T-cell receptor of $CD4^+$ T-cells. (8) The TCR-MHC-II/peptide complex activates the $CD4^+$ T-cells and orchestrates a range of immune responses, which includes stimulation of B cells for antibody production

tapasin [an ER resident chaperone, also known as TAP-binding protein (TAPBP)], ERp57 (an oxidoreductase), and calreticulin (a sugar-binding chaperone) (Margulies et al. 2020; Androlewicz et al. 1992; Neefjes et al. 1993; Kleijmeer et al. 1992; Li et al. 1999). These molecules interact with the short peptides mainly due to specific determinants found at the carboxyl termini of the peptides (Androlewicz et al. 1992; Neefjes et al. 1993; Kleijmeer et al. 1992; Li et al. 1999). In the absence of TAP proteins, the liposomal peptides are not transported to the ER or to the Golgi complex and remain in the cytoplasmic compartment (Fig. 4A, panels 1–3) (Rao et al. 1997). The peptide–lipopeptides are loaded onto empty MHC class I-β2 microglobulin molecules in the ER and then transported to the cell surface via the

Golgi complex (Fig. 2, steps 6–8). A schematic three-dimensional representative drawing of the structure of the Golgi can be found in Pagano (1990) and Krstic (1979) in which the Golgi fields of a composite Golgi apparatus were drawn from thick sections examined in a high-voltage electron microscope. Branching tubules connecting the Golgi cisternae, which create a complex network with numerous pores on the inner surface of the Golgi cisternae, can be seen in the drawing. One important post-translational modification that mainly occurs in the Golgi is glycosylation, which is driven by glycosyltransferases including sialyltransferase (Bao et al. 2015). The peptides bound to MHC class I molecules are then transported to the cell surface to interact with $CD8^+$ T-cells resulting in the activation of the T-cell and secretion of IFN$\gamma$ (Fig. 2, step 9). Since the peptide has to traverse through the Golgi complex before being transported to the cell surface, and because this is an important step in the intracellular trafficking pathway, localization of the peptide/ lipopeptide in the trans-Golgi can be visualized by the congruence of the overlaid fluorescence of the labeled protein and a vital stain for the Golgi apparatus, $C_6$NBD-ceramide, that is specific to the trans-Golgi (Lipsky and Pagano 1985; Pagano et al. 1989). The trans-Golgi is labeled with $C_6$NBD-ceramide in all of the figures presented in this article.

Although both free antigens and liposome-associated antigens (in this case, Ebola antigen) can be internalized by macrophages, as shown in Fig. 4B and c, soluble Ebola peptide remained diffuse and punctate (Fig. 4B, panel 1) while the liposome-associated Texas red-labeled Ebola glycoprotein peptide [L(TR-Ebola peptide)] was concentrated (Fig. 4C, panel 5). The Golgi staining (Fig. 4B, panel 2 and Fig. 4C, panel 6) and the antigen distributions are shown as overlays in Fig. 4B, panel 3 and Fig. 4C, panel 7, respectively (Peachman et al. 2005). Thus, unlike the liposomal antigen, soluble free antigens did not localize to the area of the trans-Golgi. Furthermore, antigen concentration within the trans-Golgi (Fig. 5B, E, H) required both the lipid and the protein components (Fig. 5D, G), since the absence of either component, the protein (Fig. 5A) or the lipid (Fig. 4B, panel 1), resulted in the absence of localization to the trans-Golgi area or to any particular region of the cell. Therefore, when the liposomes were associated with an antigen, the fluorescent lipids mainly localized in the Golgi (Fig. 5G, H); and conversely, when the liposomes were associated with an antigen that was fluorescently labeled, the antigen also colocalized in the region of the trans-Golgi (Fig. 5D, E). This pattern of intracellular trafficking of liposomal lipids was affected neither by the MHC haplotype nor by the chain length of the fatty acid in the phospholipid of the liposomes (Rao and Alving 2000). In the absence of protein in the liposomes, only 5.3% of the cells showed localization of the lipids in the Golgi. However, with liposome-associated protein, the lipids localized predominately in the Golgi in about 70% of the cells. Thus, the presence of a protein in the liposomes had a remarkable "targeting" effect on the liposomal lipid, resulting in the deposition of most of the lipid within the Golgi. Likewise, the presence of the lipid associated with the protein also resulted in the trafficking of the fluorescent peptides to the trans-Golgi (Rao and Alving 2000; Rao et al. 1999b). These results have profound implications for both protein and lipid processing pathways and presentation to

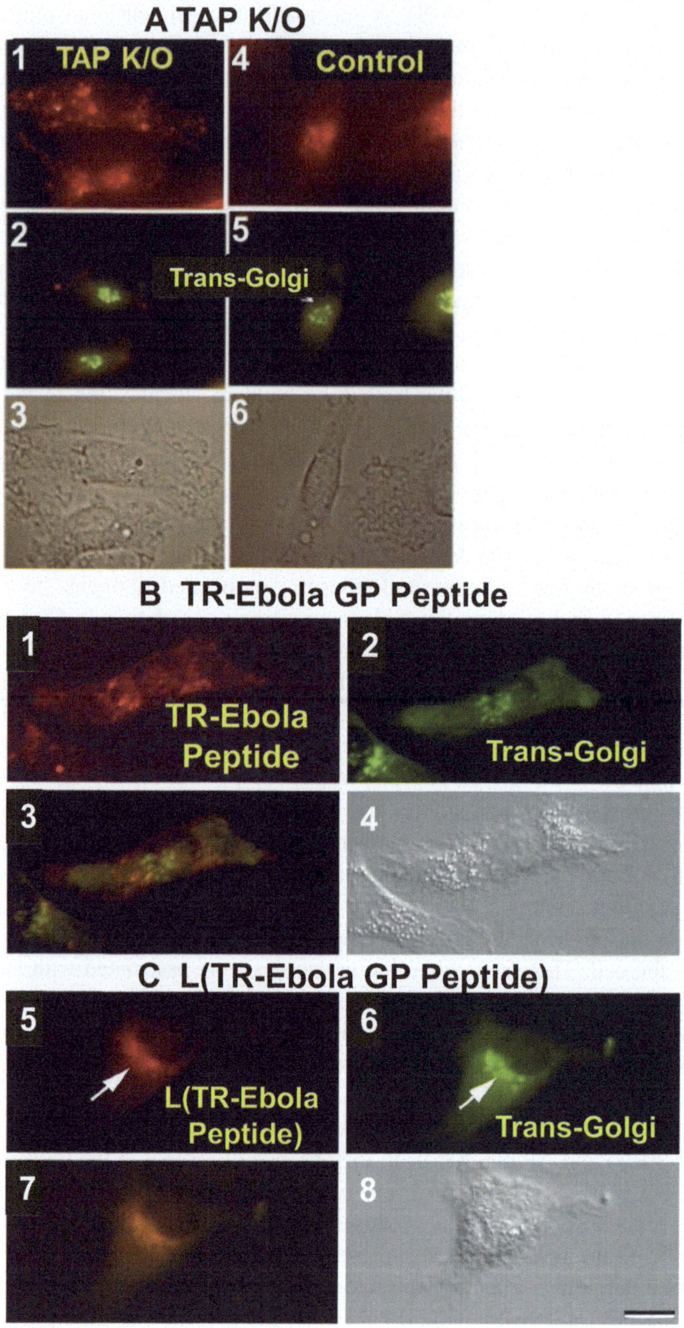
A TAP K/O
1 TAP K/O
4 Control
2
5
Trans-Golgi
3
6
B TR-Ebola GP Peptide
1
TR-Ebola
Peptide
2
Trans-Golgi
3
4
C L(TR-Ebola GP Peptide)
5
L(TR-Ebola
Peptide)
6
Trans-Golgi
7
8

◂**Fig. 4** Liposomal antigen trafficking in murine bone marrow-derived cells. **A** Importance of TAP complex. (1) Lack of concentration in the trans-Golgi of Texas red-labeled liposomal conalbumin in live bone marrow-derived macrophages from TAP1 knock-out mice. (2) Trans-Golgi was labeled with $C_6$NBD-ceramide (green color). Concentration of the labeled liposomal ovalbumin in macrophages from (4) wild-type mice in the (5) trans-Golgi. (3) and (6) The corresponding bright field images of the respective macrophages are shown. From Rao et al. (1999b). **B** Human macrophages do not concentrate soluble antigen. (1) Diffuse staining of Texas red-labeled Ebola envelope peptide in the absence of liposomes and lack of concentration in the (2) trans-Golgi labeled with $C_6$NBD-ceramide (green color) as seen with the (3) overlay image. (4) The bright field image is shown. **C** Concentration of liposomal Ebola peptide. (5) Concentration and localization of Texas red-labeled Ebola envelope peptide encapsulated in liposomes in the (6) trans-Golgi as evidenced by the (7) overlay image. The regions of complete identity appear yellowish orange. (8) The bright field image is shown. The white arrows indicate the (5) area of antigen concentration and (6) in the trans-Golgi area in panel D, respectively. Scale bar, 10 mm. From Peachman et al. (2005)

T-cells. At the present time, the intracellular pathway for lipid antigens has not been fully studied and this could be a fruitful subject for future analysis.

The peptide-MHC-I complex probably utilizes the microtubules or shuttle vesicles to reach the cell surface. The presence of this complex was visualized either by flow cytometry (Fig. 6A) or by fluorescence microscopy (Fig. 6B) with an antibody specific for the MHC class I-peptide complex, in this case the SIINFEKL complex (Porgador et al. 1997; Rothwell et al. 2000). Once on the surface, it was accessible for interaction with the TCR of $CD8^+$ T-cells (Fig. 2 step 8) (Rao et al. 1999a, b). This interaction caused cell activation and the release of cytokines such as IFN-$\gamma$ (Fig. 2 step 9). Liposome-associated protein and exogenous particulate antigens are noted for their propensities to induce IFN-$\gamma$ secretion and to induce cytotoxic T lymphocytes (CTLs) (Schmidt et al. 2019; Ramakrishnan et al. 2019a, b, c; Alving et al. 1995; Lopes and Chain 1992; Reddy et al. 1992; White et al. 1993; Rao et al. 2004). Entry into the MHC class I pathway was demonstrated by the induction of conalbumin-specific MHC class I-restricted CD8 CTL responses in mice (Fig. 6C) after intraperitoneal injection of liposomal conalbumin (Rao and Alving 2000). The data generated in vitro thus predicted the possible outcomes in vivo. Although, there are differences in the proteins required for MHC class I and MHC class II processing and presentation, antigens processed through one pathway can also be presented by the other pathway (Giodini and Albert 2010).

## 4.2 MHC Class II Processing Pathway

MHC class II molecules are expressed on professional APCs, including DCs, monocytes, macrophages, and B cells. Exogenous antigens such as liposomal antigens can enter the endosomal compartment through endocytosis, phagocytosis (Fig. 3 steps 1 and 2), or by autophagy (Munz 2010). The endosomal compartments contain numerous cathepsins and other proteases, which are cell-type specific, and

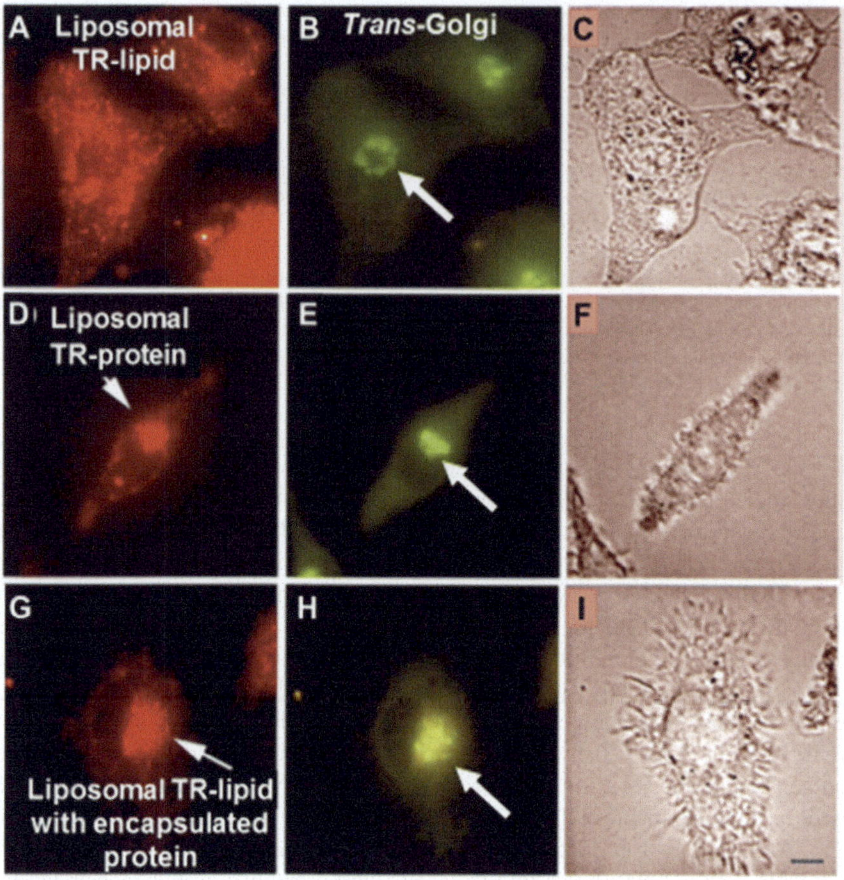


**Fig. 5** Lipid and protein requirement for Golgi-specific localization. (**A**) Diffuse staining pattern of Texas red-labeled liposomes (no protein) in live murine bone marrow-derived macrophages with a distinct absence of localization of the labeled lipid in the trans-Golgi area (**B**). The trans-Golgi (white arrow) was labeled with $C_6$NBD-ceramide (Lipsky and Pagano 1985; Pagano et al. 1989). (**D**) When the liposomes contained an antigen, Texas red-labeled conalbumin, the liposomal antigen localized to the trans-Golgi (**E**) as depicted by the white arrows. (**G**) Localization of the liposomal lipids (labeled lipids) was observed (white arrow) (**H**) When the labeled liposomes contained unlabeled antigen. Thus, both protein and lipids are essential for liposomal antigens to localize to the trans-Golgi area. The corresponding bright field images are shown respectively in (**C**, **F**, and **I**). Scale bar, 10 nm. From Rao et al. (1999b)

can be differentially expressed based on the activation status of the cell (Stern et al. 2006). Dendritic cells and nonactivated cells contain low concentrations of cathepsins, which allows for the preservation of the antigens. In contrast, macrophages and activated cells contain high concentrations of cathepsins that rapidly degrade antigens. The enzymes cleave endocytosed antigens and generate a peptide array containing both MHC class I and MHC class II epitopes. The peptides can

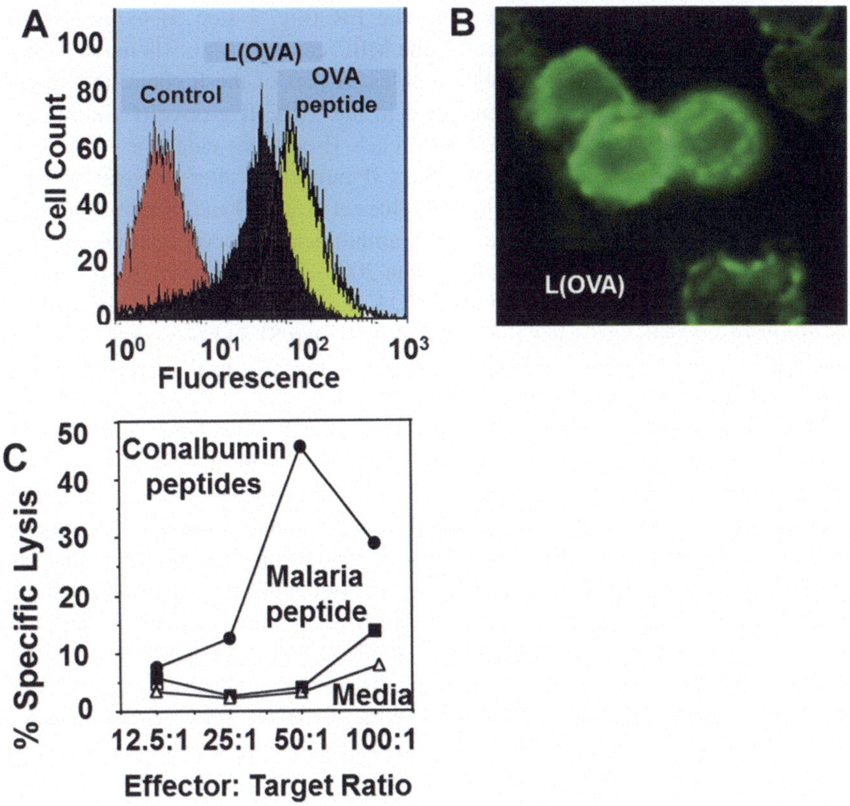


**Fig. 6** Detection of peptide-MHC class I complexes on the cell surface and induction of cytotoxic T-cells. Peptide-MHC class I complexes were detected following incubation of murine bone marrow-derived hybridoma C2.3 cells with either SIINFEKL peptide (an OVA peptide) or liposomal ovalbumin L(OVA) or buffer (control) followed by an antibody specific for the peptide–MHC complex and examined either by **A** flow cytometry or **B** fluorescence microscopy. **C** Cytotoxic T lymphocyte recall responses in murine splenic lymphocytes immunized with liposomal conalbumin, cultured with conalbumin peptides, CTL epitope of malaria CSP antigen (negative control), or media. Standard 4-hr CTL assays were performed using $^{51}$Cr-labeled peptide pulsed target cells, and percentage specific lysis was determined. From Rothwell et al. (2000), Rao and Alving (2000)

follow two routes. They are either retro-translocated into the cytosol where they can then enter the classical MHC class I pathway and be processed by the proteasome complex, or they enter the MHC class II compartment (MIIC) where they encounter the MHC class II molecules.

The α and β glycoprotein chains of the MHC class II molecule are synthesized in the ER and associate with the invariant chain (Ii) to form a heterotrimer complex. The Ii chaperone molecule stabilizes the MHC heterodimer and also prevents endogenous peptides from binding to the groove of the MHC molecule. The heterotrimer complex then transits from the ER (Fig. 3 step 3) to the cis and trans-Golgi (Fig. 3 steps 4, 5) and enters the MIIC compartment, where the Ii chain is degraded by cathepsins and other proteases. A part of the Ii chain, the class II invariant chain peptide (CLIP), is retained in the MHC-II groove (Couture et al. 2019). The peptides containing the MHC class II epitopes enter the MIIC compartment where they are loaded onto class II molecules after removal of CLIP (Fig. 3 step 6) (Roche and Cresswell 1991; Boes et al. 2005) and then transported to the cell surface (Fig. 3 step 7) as a 12–15 amino acid epitope bound to an MHC class II molecule (Stern et al. 2006; Chapman 2006). The peptide-MHC-II complex then interacts with the T-cell receptor on $CD4^+$ T-cells and in conjunction with the costimulatory signals, activates the T-cell (Fig. 3 step 8) and orchestrates a broad range of immune responses, which include providing help to the B cells for antibody production (Rao et al. 2018).

HIV-1 Env-A244 gp120 protein, one of the protein boost subunits of the RV144 phase 3 vaccine trial, was shown to be resistant to proteasomes but susceptible to cathepsin cleavage, and the resulting peptides were identified by mass spectrometry. The peptide fragments induced a polyfunctional cytokine response including the generation of IFN-$\gamma$ from $CD4^+$ T-cell lines derived from RV144 vaccinated volunteers (Steers et al. 2012). In vivo studies with both liposome-encapsulated as well as liposome-associated antigens have demonstrated robust induction of antibody and cellular immune responses (Rao et al. 2018; Cawlfield et al. 2019; Ramakrishnan et al. 2019a, b, c; Beck et al. 2015b, 2018; Genito et al. 2017; Seth et al. 2017; Alving et al. 2016; Wieczorek et al. 2015) thus proving that liposomal antigens can enter both MHC class I and class II pathways in vitro and in vivo.

As shown above, anionic liposomes containing MPLA undergo a complex choreographed intracellular trafficking pattern in phagocytes leading to generation of Th1-type and Th2-type immunity. It is important to note that ALF-like liposomes lacking MPLA often have little or no immunostimulant adjuvant activity (Alving et al. 1986; White et al. 1995). Furthermore, the liposomal lipid serves as a physical chaperone to deliver the processed antigen through the Golgi to present the peptide together with MHC class I and class II molecules on the cell surface, a trafficking pattern that is not available to the antigen in the absence of the liposome.

## 5 Rational Selection of Liposomal Human Vaccine Adjuvant Formulations

Immunization of human volunteers with experimental vaccine formulations has long been viewed as a benchmark for evaluation and rational selection of constituents likely to achieve scientific and commercial success (Alving 2002). Certainly, for analysis of the potential for toxicity, feedback from human recipients is a major requirement; and challenge of human volunteers with the target pathogen can lead to even more rigorous analysis. Obviously, caution is required in any type of human immunization, and animal models are often critically important both for preclinical evaluation of potential vaccine efficacy and toxicity. However, in the absence of a rigorous animal model of the human disease, such as is the case for a therapeutic vaccine to prostate cancer which is unique to humans, human experimentation is required (Alving 2002).

Adjuvant research, and particularly malaria vaccine adjuvant research, has been a major beneficiary of the availability of human challenge models, such as the controlled human malaria infection (CHMI) for which curative drugs are available (Chulay et al. 1986). In 1984, the gene encoding the circumsporozoite (CS) protein on the outer surface of *P. falciparum* was cloned and sequenced (Dame et al. 1984; Enea et al. 1984). Of particular importance for vaccine development was a region in the middle of the protein composed of 37 Asn-Ala-Asn-Pro tetrapeptide repeats interspersed with four Asn-Val-Asp-Pro tetrapeptides. Although murine antibodies to recombinant proteins comprising multiple copies of the tetrapeptides recognized the complete CS protein, higher titers of the antibodies in experimental animals required the presence of either aluminum hydroxide or Freund's adjuvant (Young et al. 1985). In the first phase 1 human trial of a recombinant protein (R32tet32) adsorbed to aluminum hydroxide (falciparum sporozoite vaccine-1; known as FSV-1), six immunized volunteers were challenged by CHMI with the bite of infected mosquitoes. The good news was that protection was observed against infection; but the disappointing news was that fifty weeks after the first immunization, and after three subsequent boosts of vaccine, only 1 of 6 volunteers was protected, and protection was observed only in the volunteer with the highest antibody titer (Ballou et al. 1987; Young et al. 1987).

At the same time that the FSV-1 malaria vaccine was being developed, a new adjuvant, known as Walter Reed liposomes, and later known as Army Liposome Formulation (ALF), was being developed at WRAIR for additional human vaccine trials (Alving et al. 1986, 2020). Based on the previous disappointing results observed with aluminum-adsorbed FSV-1, the first use of a liposome-based vaccine in humans was undertaken, and a recombinant protein containing the CS protein tetrapeptides was used as the antigen (Fries et al. 1992). In this study, two adjuvants were present in the vaccine formulation: MPLA in the liposomal bilayer, and the liposomes were also adsorbed to aluminum hydroxide as a second adjuvant. In addition to strong immune potency, the liposomes completely blocked the inherent pyrogenicity of MPLA, the primary signal of toxicity, even at potentially lethal amounts (>2 mg of MPLA). When compared to the FSV-1 vaccine, which contained only aluminum salt as an adjuvant, anti-NANP titers induced by the aluminum-adsorbed liposomal MPLA formulation were tenfold higher. When later compared with six other adjuvants in a multicenter phase 1 trial with a single antigen (recombinant gp120 envelope protein from HIV-1) (AVEG 015), the aluminum-adsorbed ALF formulation containing encapsulated antigen had greater apparent potency and the lowest apparent toxicity (equivalent to aluminum hydroxide alone) (Mcelrath 1995). Recent analysis of sera obtained from individuals in the AVEG 015 trial immunized with aluminum-adsorbed antigen or aluminum-adsorbed ALF containing encapsulated antigen revealed dramatically increased anti-gp120 titers and increased antibody duration, and high titers of antibodies specific to V1/V2 loop of the gp120 envelope protein (Rao et al. 2018).

In our experience, the variables to be considered in construction of liposomal formulations are many, but first and foremost the ability to sterilize the preparation is critical. This can be achieved by sterile filtration of each of the components at some point during the manufacture through a 0.2-μm filter and maintaining sterility afterward (Matyas et al. 2003). Thus, liposomes that are nanoparticles can be easily sterile-filtered after manufacture. However, for anionic liposomes the requirement for nanosized particles potentially limits the use of phospholipids to those that have unsaturated fatty acids which automatically form nanoliposomes. Liposomes with saturated phospholipids (such as ALF) can be manufactured by the ethanol injection technique (Wagner and Vorauer-Uhl 2011), or by the use of microfluidization to create nanosized particles. The chemical composition of the liposomes, the physical chemistry of the liposomal membrane scaffold, and the type and mechanism of action of the liposome-associated adjuvant, are all important factors. For example, in studies in which the protective antigen (PA) of anthrax was adjuvanted with aluminum salt (as in the licensed anthrax vaccine) and then improved by using ALF instead of aluminum salt, it was discovered that liposome encapsulation of PA was not necessary, and the PA could simply be mixed with ALF in formulating the vaccine (Rao et al. 2011).

In the above-described first human liposomal vaccine trial with malaria antigen, and in the AVEG 015 trial with HIV-1 gp120, the liposomal adjuvant formulation (ALF) was manufactured at WRAIR using synthetic phospholipids and natural MPLA (MPL®) extracted from Salmonella bacterial lipopolysaccharide supplied by Ribi ImmunoChem Research, Inc. The liposomes contained a heterogeneous population of large (micro) and small (nano) liposomes, thus preventing sterile filtration with a 0.22 μm filter; and in this case, the liposomes required sterile manufacture. Because of the potency of the Ribi MPLA as a human vaccine adjuvant, the Ribi company became valuable as an adjuvant manufacturing company, and it now belongs to GSK. The ALF-type adjuvant was re-worked by GSK and expanded to include a formulation of nanoliposomes containing both MPLA and the saponin QS21 as adjuvants, which is now known as Adjuvant System 01 (AS01) (Vandepapeliere 2018; Alving et al. 2020). After numerous comparative clinical trials, a successful sporozoite malaria vaccine (RTS,S/AS01, also known as Mosquirix®) with a recombinant protein containing the repeat tetrapeptides of the CS protein (RTS,S) and the AS01 adjuvant emerged. The formulation received a positive scientific opinion by the European Medicines Agency exclusively for markets outside of the European Union and is currently being further tested with a four-injection regimen under World Health Organization auspices among hundreds of thousands of children 5–9 months of age in settings of moderate-to-high parasite transmission for malaria in sub-Saharan Africa (World Health Organization 2018). The AS01 adjuvant has also enabled a highly successful vaccine to Herpes zoster (Shingrix®) that is currently licensed for human use by the U.S. Food and Drug Administration for prevention of shingles (Lal et al. 2015).

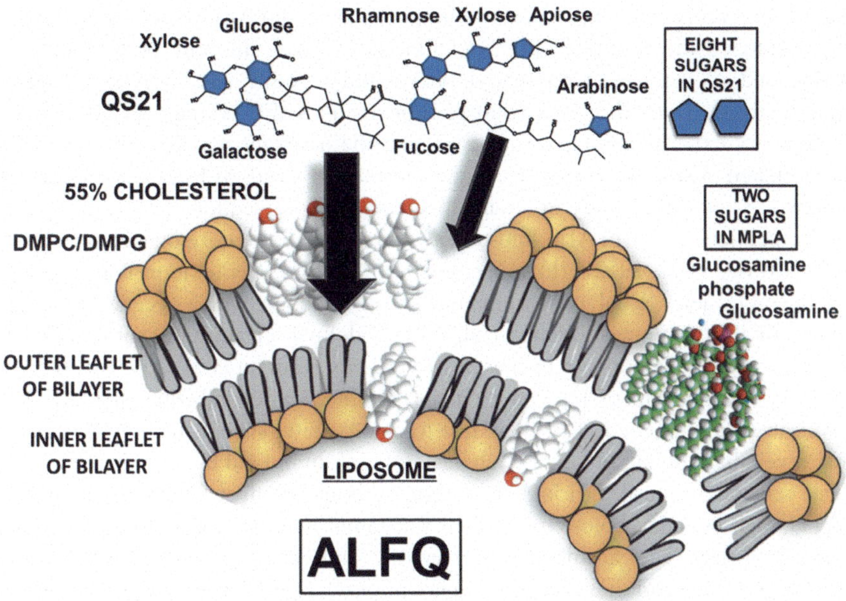


**Fig. 7** Schematic illustration of the Army Liposome Formulation containing QS21 (ALFQ). Liposomes containing 55% cholesterol and monophosphoryl lipid A are shown with QS-21 which is approaching the cholesterol where it binds tightly via hydrophobic regions. This enables the eight sugars of QS21 to be displayed at the liposomal surface. Monophosphoryl lipid A adds two more sugars at the water interface, thus forming a complex and diverse sugar display (a "sugar lawn") by ALFQ on the liposome surface. From Alving et al. (2020)

In view of the above sequence of events, it seemed reasonable to presume that the AS01 adjuvant could serve as a rational model for further development of similar types of liposomal vaccine adjuvant formulations (Alving et al. 2020). To that end, ALFQ, namely ALF liposomes containing QS21 saponin, was created as a new entry in the ALF family of adjuvants (Beck et al. 2015a, b) (Fig. 7). ALFQ was created by adding QS21 to microfluidized nanoliposomes (50–100 nm diameter), thus allowing sterile filtration prior to this step. However, in contrast to the AS01 formulation, addition of sterile-filtered QS21 to the nanoliposomes caused a huge increase in the size of the liposomes, to produce large unilamellar or oligolamellar vesicles, as large as 30,000 nm in diameter (Beck et al. 2015b). This size change occurred through cannibalization of the nanoparticles to form microparticles. Although some nanoparticles remained, they were greatly reduced in number, and the vast majority of the total volume and total surface area of the ALFQ particles were likely contributed by microparticles instead of nanoparticles. To illustrate this by solid geometry, a single 30,000 nm spherical particle contains a volume that is >216 million times greater than that of a 50 nm particle, and the surface area of the single outer leaflet of the lipid bilayer of a 30,000 nm particle is approximately 360,000 times greater than that of a 50 nm particle. Further investigation revealed that the difference in the liposome size between AS01 and ALFQ was due to the use of unsaturated dioleoyl phosphatidylcholine (DOPC) as the bulk phospholipid in AS01, and of saturated dimyristoyl phosphatidylcholine and dimyristoyl phosphatidylglycerol as the bulk phospholipids in ALFQ (Singh et al. 2019). Preliminary experiments have suggested that the adjuvant potencies in mice of a lab-prepared AS01-like formulation and the micron-sized ALFQ liposome preparations are not significantly different (manuscript in preparation).

As mentioned above, ALF liposomes completely lack the pyrogenic toxicity of MPLA (Fries et al. 1992). However, free QS21 by itself is toxic in that it causes considerable acute local pain immediately on injection which recedes over the course of thirty minutes (Waite et al. 2001; Edelman et al. 2002), and it causes necrosis at the injection site in muscles in rodents (Garcon et al. 2007). Although the exact mechanism of the acute pain experienced after injection of QS21 has never been fully studied, it is likely due to cytotoxicity, such as hemolysis, caused by irreversible binding of QS21 to cholesterol present in lipid bilayer membranes of local cells (Beck et al. 2015a). To mitigate this toxic effect of QS21, AS01 and ALFQ each contain cholesterol in their liposomal membranes, and the liposomal cholesterol serves as a sink that irreversibly and strongly binds to QS21, and liposomal cholesterol thus blocks the binding of QS21 to cells in the vicinity of the injection site. Still, despite the absence of immediate pain, many who have received the Shingrix® vaccine that contains AS01B as an adjuvant have experienced local reactions at the injection site that occur hours after injection and which last for several days. If these delayed local effects are due to the AS01 adjuvant, it might seem unlikely that they are caused by the QS21 because of the strong and irreversible binding of QS21 to the liposomal cholesterol; and if this is so, then either the DOPC or the MPLA in the AS01 might somehow be the culprit. It is conceivable that MPLA that is present in AS01 might be expressing local toxicity due to incomplete blocking by the highly unsaturated phospholipid environment of AS01. Two ongoing clinical trials utilizing ALFQ adjuvant with two different malaria vaccines may help to determine whether reactogenicity is experienced by humans after injections of vaccines containing ALFQ (Alving et al. 2020).

Regardless of the issues of reactogenicity and how to address them, the mechanisms underlying the extraordinary potency of AS01 as an adjuvant for Mosquirix® and Shingrix® are still subjects of ongoing research, but it seems clear that the combined constituents of AS01, namely DOPC, cholesterol, MPLA, and QS21, mobilizes multiple mechanisms of innate immunity leading to the adaptive immune response (Alving et al. 2012; Didierlaurent et al. 2014; Welsby et al. 2016). Nanoliposomes are clearly important as a practical means for achieving sterility of liposomal vaccine development; however, it should be noted that although ALFQ starts with nanoliposomes, QS21 interacting with the high concentrations of cholesterol transforms the vesicles from nanosize to microsize (Singh et al. 2019; Alving and Beck 2019). ALFQ is easily manufactured by using sterile-filtered QS21 that is added to the sterile-filtered ALF nanoliposomes. In summary, adjuvant research is still a dynamic and developing field in which successful adjuvant products, including liposomal adjuvants, that are present in licensed human vaccines provide important directions for rational adjuvant selection.

## 6 Conclusion

Early discoveries of liposome–protein complexes (such as complement protein complexes) emphasized the impact of exposure of liposomes to biological proteins such as complement on the delivery and uptake of liposome-associated antigens by APCs. This has been further informed by research from the nanomedicines field on the importance of a so-called protein corona of proteins attached to nanoparticles introduced to the biosphere. While the protein corona of a particle sometimes may have a detrimental effect on delivery of nanomedicines, in the vaccine adjuvant field a protein corona attached to liposomes can be an important factor for delivery of liposomal antigens to APCs. The intracellular trafficking patterns of liposomes and liposome-associated protein in APCs provided important insights into the mechanisms by which liposomes can influence the immune response. For example, these studies indicated that liposome-associated peptide, but not free peptide, can transit through the trans-Golgi, and a complex structure of lipid and protein thus appears to be an important element for the MHC class I processing pathway in phagocytic cells.

Because of the continuous fifty-year background of experience with liposome-associated proteins and different types of liposomes at WRAIR, there have been smooth transitions to translational applications (Fig. 8). The

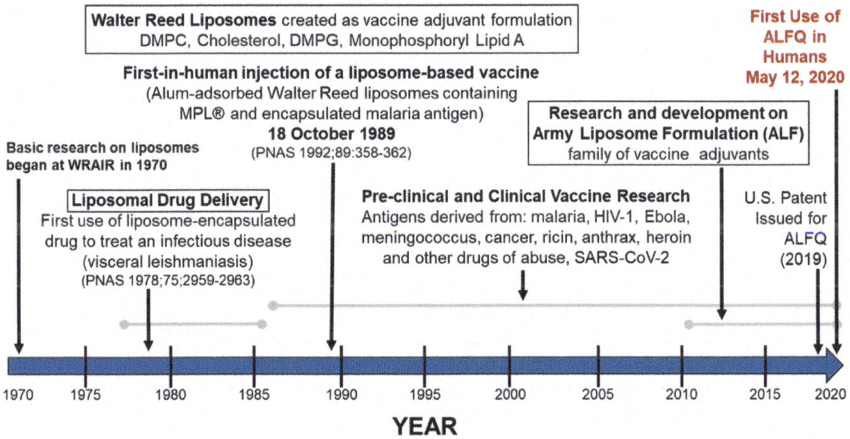


**Fig. 8** Selected milestones of liposome research at WRAIR. Liposome research has been ongoing at WRAIR for fifty years, since 1970. It has been undertaken with three foci: **a** basic membrane biochemistry, cellular interactions, and immunological research; **b** drug delivery research; and **c** research on liposomes as adjuvants for many types of vaccines

development of the Army Liposome Formulation (ALF) family of vaccine adjuvants is currently being employed in a variety of phase 1 studies with different types of candidate vaccines. As with the first liposome vaccine injected into humans in 1982, which comprised a candidate malaria vaccine, the first injections of candidate vaccines containing ALFQ as antigens are currently ongoing with two different types of candidate malaria vaccines (Alving et al. 2020). The overall safety and potency of the vaccines containing ALFQ will be evaluated over the next year.

**Disclaimer** The views expressed are those of the authors and should not be construed to represent the positions of the US Army or the Department of Defense.

## References


Allison AG, Gregoriadis G (1974) Liposomes as immunological adjuvants. Nature 252:252

Alving CR (1984) Natural antibodies against phospholipids and liposomes in humans. Biochem Soc Trans 12:342–344

Alving CR (1991) Liposomes as carriers of antigens and adjuvants. J Immunol Methods 140:1–13

Alving CR (2002) Design and selection of vaccine adjuvants: animal models and human trials. Vaccine 20(Suppl 3):S56–S64

Alving C, Beck Z (2019) Non-toxic adjuvant formulation comprising a monophosphoryl lipid A (MPLA)-containing liposome composition and a saponin. US patent application, 8 Oct 2019

Alving CR, Kinsky SC (1971) The preparation and properties of liposomes in the LA and LAC states. Immunochemistry 8:325–343

Alving CR, Wassef NM (1999) Naturally occurring antibodies to cholesterol: a new theory of LDL cholesterol metabolism. Immunol Today 20:362–366

Alving CR, Richards RL, Moss J et al (1986) Effectiveness of liposomes as potential carriers of vaccines: applications to cholera toxin and human malaria sporozoite antigen. Vaccine 4:166–172

Alving CR, Koulchin V, Glenn GM et al (1995) Liposomes as carriers of peptide antigens: induction of antibodies and cytotoxic T lymphocytes to conjugated and unconjugated peptides. Immunol Rev 145:5–31

Alving CR, Peachman KK, Rao M et al (2012) Adjuvants for human vaccines. Curr Opin Immunol 24:310–315

Alving CR, Beck Z, Matyas GR et al (2016) Liposomal adjuvants for human vaccines. Expert Opin Drug Deliv 13:807–816

Alving CR, Peachman KK, Matyas GR et al (2020) Army Liposome Formulation (ALF) family of vaccine adjuvants. Expert Rev Vaccines 1–14

Androlewicz MJ, Browning JL, Ware CF (1992) Lymphotoxin is expressed as a heteromeric complex with a distinct 33-kDa glycoprotein on the surface of an activated human T cell hybridoma. J Biol Chem 267:2542–2547

Ballou WR, Hoffman SL, Sherwood JA et al (1987) Safety and efficacy of a recombinant DNA *Plasmodium falciparum* sporozoite vaccine. Lancet 1:1277–1281

Bangham AD, Standish MM, Watkins JC (1965) Diffusion of univalent ions across the lamellae of swollen phospholipids. J Mol Biol 13:238–252

Bao L, Ding L, Yang M et al (2015) Noninvasive imaging of sialyltransferase activity in living cells by chemoselective recognition. Sci Rep 5:10947

Beck Z, Matyas GR, Alving CR (2015a) Detection of liposomal cholesterol and monophosphoryl lipid A by QS-21 saponin and Limulus polyphemus amebocyte lysate. Biochim Biophys Acta 1848:775–780

Beck Z, Matyas GR, Jalah R et al (2015b) Differential immune responses to HIV-1 envelope protein induced by liposomal adjuvant formulations containing monophosphoryl lipid A with or without QS21. Vaccine 33:5578–5587

Beck Z, Torres OB, Matyas GR et al (2018) Immune response to antigen adsorbed to aluminum hydroxide particles: effects of co-adsorption of ALF or ALFQ adjuvant to the aluminum-antigen complex. J Control Release 275:12–19

Bevan MJ (1987) Antigen recognition. Class discrimination in the world of immunology. Nature 325:192–194

Boes M, Van Der Wel N, Peperzak V et al (2005) In vivo control of endosomal architecture by class II-associated invariant chain and cathepsin S. Eur J Immunol 35:2552–2562

Bohlson SS, Garred P, Kemper C et al (2019) Complement nomenclature-deconvoluted. Front Immunol 10:1308

Bonte F, Juliano RL (1986) Interactions of liposomes with serum proteins. Chem Phys Lipids 40:359–372

Cawlfield A, Genito CJ, Beck Z et al (2019) Safety, toxicity and immunogenicity of a malaria vaccine based on the circumsporozoite protein (FMP013) with the adjuvant army liposome formulation containing QS21 (ALFQ). Vaccine 37:3793–3803

Cedervall T, Lynch I, Lindman S et al (2007) Understanding the nanoparticle-protein corona using methods to quantify exchange rates and affinities of proteins for nanoparticles. Proc Natl Acad Sci U S A 104:2050–2055

Chapman HA (2006) Endosomal proteases in antigen presentation. Curr Opin Immunol 18:78–84

Chonn A, Cullis PR, Devine DV (1991) The role of surface charge in the activation of the classical and alternative pathways of complement by liposomes. J Immunol 146:4234–4241

Chonn A, Semple SC, Cullis PR (1992) Association of blood proteins with large unilamellar liposomes in vivo. Relation to circulation lifetimes. J Biol Chem 267:18759–18765

Chulay JD, Schneider I, Cosgriff TM et al (1986) Malaria transmitted to humans by mosquitoes infected from cultured *Plasmodium falciparum*. Am J Trop Med Hyg 35:66–68

Collins DS, Findlay K, Harding CV (1992) Processing of exogenous liposome-encapsulated antigens in vivo generates class I MHC-restricted T cell responses. J Immunol 148:3336–3341

Couture A, Garnier A, Docagne F et al (2019) HLA-Class II artificial antigen presenting cells in CD4(+) T cell-based immunotherapy. Front Immunol 10:1081

Dame JB, Williams JL, Mccutchan TF et al (1984) Structure of the gene encoding the immunodominant surface antigen on the sporozoite of the human malaria parasite *Plasmodium falciparum*. Science 225:593–599

De Serrano LO, Burkhart DJ (2017) Liposomal vaccine formulations as prophylactic agents: design considerations for modern vaccines. J Nanobiotechnol 15:83

Didierlaurent AM, Collignon C, Bourguignon P et al (2014) Enhancement of adaptive immunity by the human vaccine adjuvant AS01 depends on activated dendritic cells. J Immunol 193:1920–1930

Didierlaurent AM, Laupeze B, Di Pasquale A et al (2017) Adjuvant system AS01: helping to overcome the challenges of modern vaccines. Expert Rev Vaccines 16:55–63

Edelman R, Wasserman SS, Kublin JG et al (2002) Immediate-type hypersensitivity and other clinical reactions in volunteers immunized with a synthetic multi-antigen peptide vaccine (PfCS-MAP1NYU) against Plasmodium falciparum sporozoites. Vaccine 21:269–280

Enea V, Ellis J, Zavala F et al (1984) DNA cloning of Plasmodium falciparum circumsporozoite gene: amino acid sequence of repetitive epitope. Science 225:628–630

Engler OB, Schwendener RA, Dai WJ et al (2004) A liposomal peptide vaccine inducing CD8+ T cells in HLA-A2.1 transgenic mice, which recognise human cells encoding hepatitis C virus (HCV) proteins. Vaccine 23:58–68

Fries LF, Gordon DM, Richards RL et al (1992) Liposomal malaria vaccine in humans: a safe and potent adjuvant strategy. Proc Natl Acad Sci U S A 89:358–362

Garcon N, Chomez P, Van Mechelen M (2007) GlaxoSmithKline adjuvant systems in vaccines: concepts, achievements and perspectives. Expert Rev Vaccines 6:723–739

Genito CJ, Beck Z, Phares TW et al (2017) Liposomes containing monophosphoryl lipid A and QS-21 serve as an effective adjuvant for soluble circumsporozoite protein malaria vaccine FMP013. Vaccine 35:3865–3874

Germain RN, Margulies DH (1993) The biochemistry and cell biology of antigen processing and presentation. Annu Rev Immunol 11:403–450

Giodini A, Albert ML (2010) A whodunit: an appointment with death. Curr Opin Immunol 22:94–108

Gregoriadis G (1978) Liposomes in therapeutic and preventive medicine: the development of the drug-carrier concept. Ann N Y Acad Sci 308:343–370

Haxby JA, Kinsky CB, Kinsky SC (1968) Immune response of a liposomal model membrane. Proc Natl Acad Sci U S A 61:300–307

Haxby JA, Gotze O, Muller-Eberhard HJ et al (1969) Release of trapped marker from liposomes by the action of purified complement components. Proc Natl Acad Sci U S A 64:290–295

Henriksen-Lacey M, Bramwell VW, Christensen D et al (2010) Liposomes based on dimethyldioctadecylammonium promote a depot effect and enhance immunogenicity of soluble antigen. J Control Release 142:180–186

Heppner DG, Gordon DM, Gross M et al (1996) Safety, immunogenicity, and efficacy of Plasmodium falciparum repeatless circumsporozoite protein vaccine encapsulated in liposomes. J Infect Dis 174:361–366

Herzog C, Hartmann K, Kunzi V et al (2009) Eleven years of inflexal V-a virosomal adjuvanted influenza vaccine. Vaccine 27:4381–4387

Jafari I, Heravi Shargh V, Shahryari M et al (2018) Cationic liposomes formulated with a novel whole Leishmania lysate (WLL) as a vaccine for leishmaniasis in murine model. Immunobiology 223:493–500

Joffre OP, Segura E, Savina A et al (2012) Cross-presentation by dendritic cells. Nat Rev Immunol 12:557–569

Kamphuis T, Meijerhof T, Stegmann T et al (2012) Immunogenicity and protective capacity of a virosomal respiratory syncytial virus vaccine adjuvanted with monophosphoryl lipid A in mice. PLoS ONE 7:e36812

Ke PC, Lin S, Parak WJ et al (2017) A decade of the protein corona. ACS Nano 11:11773–11776

Kinsky SC, Haxby JA, Zopf DA et al (1969) Complement-dependent damage to liposomes prepared from pure lipids and *Forssman hapten*. Biochemistry 8:4149–4158

Kleijmeer MJ, Kelly A, Geuze HJ et al (1992) Location of MHC-encoded transporters in the endoplasmic reticulum and cis-Golgi. Nature 357:342–344

Krstic RE (1979) Ultrastructure of the Mammalian Cell. *An Atlas*. Springer-Verlag, New York

Laborde RJ, Sanchez-Ferras O, Luzardo MC et al (2017) Novel adjuvant based on the pore-forming protein sticholysin II encapsulated into liposomes effectively enhances the antigen-specific CTL-mediated immune response. J Immunol 198:2772–2784

Lal H, Cunningham AL, Godeaux O et al (2015) Efficacy of an adjuvanted herpes zoster subunit vaccine in older adults. N Engl J Med 372:2087–2096

Li S, Paulsson KM, Sjogren HO et al (1999) Peptide-bound major histocompatibility complex class I molecules associate with tapasin before dissociation from transporter associated with antigen processing. J Biol Chem 274:8649–8654

Li M, Du C, Guo N et al (2019) Composition design and medical application of liposomes. Eur J Med Chem 164:640–653

Liang F, Lore K (2016) Local innate immune responses in the vaccine adjuvant-injected muscle. Clin Transl Immunology 5:e74

Liang F, Ploquin A, Hernandez JD et al (2015) Dissociation of skeletal muscle for flow cytometric characterization of immune cells in macaques. J Immunol Methods 425:69–78

Lipsky NG, Pagano RE (1985) A vital stain for the Golgi apparatus. Science 228:745–747

Liszewski MK, Elvington M, Kulkarni HS et al (2017) Complement's hidden arsenal: new insights and novel functions inside the cell. Mol Immunol 84:2–9

Lopes LM, Chain BM (1992) Liposome-mediated delivery stimulates a class I-restricted cytotoxic T cell response to soluble antigen. Eur J Immunol 22:287–290

Lubbers R, Van Essen MF, Van Kooten C et al (2017) Production of complement components by cells of the immune system. Clin Exp Immunol 188:183–194

Margulies DH, Jiang J, Natarajan K (2020) Structural and dynamic studies of TAPBPR and Tapasin reveal the mechanism of peptide loading of MHC-I molecules. Curr Opin Immunol 64:71–79

Matyas GR, Alving CR (2011) Antigen-specific enhancement of natural human IgG antibodies to phosphatidylcholine, phosphatidylglycerol, phosphatidylinositol-4-phosphate, cholesterol, and lipid A by a liposomal vaccine containing lipid A. Vaccine 29:5137–5144

Matyas GR, Muderhwa JM, Alving CR (2003) Oil-in-water liposomal emulsions for vaccine delivery. Methods Enzymol 373:34–50

Mcelrath MJ (1995) Selection of potent immunological adjuvants for vaccine construction. Semin Cancer Biol 6:375–385

Moser C, Muller M, Kaeser MD et al (2013) Influenza virosomes as vaccine adjuvant and carrier system. Expert Rev Vaccines 12:779–791

Munz C (2010) Antigen processing via autophagy–not only for MHC class II presentation anymore? Curr Opin Immunol 22:89–93

Neefjes JJ, Momburg F, Hammerling GJ (1993) Selective and ATP-dependent translocation of peptides by the MHC-encoded transporter. Science 261:769–771

Nelson RA Jr, Jensen J, Gigli I et al (1966) Methods for the separation, purification and measurement of nine components of hemolytic complement in guinea-pig serum. Immunochemistry 3:111–135

Nisini R, Poerio N, Mariotti S et al (2018) The multirole of liposomes in therapy and prevention of infectious diseases. Front Immunol 9:155

Nunes-Hasler P, Demaurex N (2017) The ER phagosome connection in the era of membrane contact sites. Biochim Biophys Acta Mol Cell Res 1864:1513–1524

Pagano RE (1990) The Golgi apparatus: insights from lipid biochemistry. Biochem Soc Trans 18:361–366

Pagano RE, Sepanski MA, Martin OC (1989) Molecular trapping of a fluorescent ceramide analogue at the Golgi apparatus of fixed cells: interaction with endogenous lipids provides a trans-Golgi marker for both light and electron microscopy. J Cell Biol 109:2067–2079

Peachman KK, Rao M, Palmer DR et al (2004) Functional microtubules are required for antigen processing by macrophages and dendritic cells. Immunol Lett 95:13–24

Peachman KK, Rao M, Alving CR et al (2005) Human dendritic cells and macrophages exhibit different intracellular processing pathways for soluble and liposome-encapsulated antigens. Immunobiology 210:321–333

Pedersen GK, Andersen P, Christensen D (2018) Immunocorrelates of CAF family adjuvants. Semin Immunol 39:4–13

Perry DG, Martin WJ 2nd (1995) Fluorescent liposomes as quantitative markers of phagocytosis by alveolar macrophages. J Immunol Methods 181:269–285

Pick H, Alves AC, Vogel H (2018) Single-vesicle assays using liposomes and cell-derived vesicles: from modeling complex membrane processes to synthetic biology and biomedical applications. Chem Rev 118:8598–8654

Porgador A, Yewdell JW, Deng Y et al (1997) Localization, quantitation, and in situ detection of specific peptide-MHC class I complexes using a monoclonal antibody. Immunity 6:715–726

Ramakrishnan A, Schumack NM, Gariepy CL et al (2019a) Correction for Ramakrishnan et al., "Enhanced immunogenicity and protective efficacy of a *Campylobacter jejuni* conjugate vaccine coadministered with liposomes containing monophosphoryl lipid A and QS-21". mSphere 4

Ramakrishnan A, Schumack NM, Gariepy CL et al (2019b) Enhanced immunogenicity and protective efficacy of a *Campylobacter jejuni* conjugate vaccine coadministered with liposomes containing monophosphoryl lipid A and QS-21. mSphere 4

Ramakrishnan A, Schumack NM, Gariepy CL et al (2019c) Erratum for Ramakrishnan et al., "Enhanced immunogenicity and protective efficacy of a *Campylobacter jejuni* conjugate vaccine coadministered with liposomes containing monophosphoryl lipid A and QS-21". mSphere 4

Rao M, Alving CR (2000) Delivery of lipids and liposomal proteins to the cytoplasm and Golgi of antigen-presenting cells. mangala.rao@na.amedd.army.mil. Adv Drug Deliv Rev 41:171–188

Rao M, Rothwell SW, Wassef NM et al (1997) Visualization of peptides derived from liposome-encapsulated proteins in the trans-Golgi area of macrophages. Immunol Lett 59:99–105

Rao M, Matyas GR, Grieder F et al (1999a) Cytotoxic T lymphocytes to Ebola Zaire virus are induced in mice by immunization with liposomes containing lipid A. Vaccine 17:2991–2998

Rao M, Rothwell SW, Wassef NM et al (1999b) Trafficking of liposomal antigen to the trans-Golgi of murine macrophages requires both liposomal lipid and liposomal protein. Exp Cell Res 246:203–211

Rao M, Bray M, Alving CR et al (2002) Induction of immune responses in mice and monkeys to Ebola virus after immunization with liposome-encapsulated irradiated Ebola virus: protection in mice requires CD4(+) T cells. J Virol 76:9176–9185

Rao M, Matyas GR, Vancott TC et al (2004) Immunostimulatory CpG motifs induce CTL responses to HIV type I oligomeric gp140 envelope protein. Immunol Cell Biol 82:523–530

Rao M, Peachman KK, Li Q et al (2011) Highly effective generic adjuvant systems for orphan or poverty-related vaccines. Vaccine 29:873–877

Rao M, Onkar S, Peachman KK et al (2018) Liposome-encapsulated human immunodeficiency virus-1 gp120 induces potent V1V2-specific antibodies in humans. J Infect Dis 218:1541–1550

Reddy R, Zhou F, Huang L et al (1991) pH sensitive liposomes provide an efficient means of sensitizing target cells to class I restricted CTL recognition of a soluble protein. J Immunol Methods 141:157–163

Reddy R, Zhou F, Nair S et al (1992) In vivo cytotoxic T lymphocyte induction with soluble proteins administered in liposomes. J Immunol 148:1585–1589

Roche PA, Cresswell P (1991) Proteolysis of the class II-associated invariant chain generates a peptide binding site in intracellular HLA-DR molecules. Proc Natl Acad Sci U S A 88:3150–3154

Rothwell SW, Wassef NM, Alving CR et al (2000) Proteasome inhibitors block the entry of liposome-encapsulated antigens into the classical MHC class I pathway. Immunol Lett 74:141–152

Schmidt ST, Olsen CL, Franzyk H et al (2019) Comparison of two different PEGylation strategies for the liposomal adjuvant CAF09: towards induction of CTL responses upon subcutaneous vaccine administration. Eur J Pharm Biopharm 140:29–39

Schwendener RA (2014) Liposomes as vaccine delivery systems: a review of the recent advances. Ther Adv Vaccines 2:159–182

Segal AW, Wills EJ, Richmond JE et al (1974) Morphological observations on the cellular and subcellular destination of intravenously administered liposomes. Br J Exp Pathol 55:320–327

Seth L, Bingham Ferlez KM, Kaba SA et al (2017) Development of a self-assembling protein nanoparticle vaccine targeting *Plasmodium falciparum* circumsporozoite protein delivered in three army liposome formulation adjuvants. Vaccine 35:5448–5454

Singh P, Beck Z, Matyas GR et al (2019) Saturated phospholipids are required for nano- to micron-size transformation of cholesterol-containing liposomes upon QS21 addition. J Liposome Res 29:247–250

Smith GE, Flyer DC, Raghunandan R et al (2013) Development of influenza H7N9 virus like particle (VLP) vaccine: homologous A/Anhui/1/2013 (H7N9) protection and heterologous A/chicken/Jalisco/CPA1/2012 (H7N3) cross-protection in vaccinated mice challenged with H7N9 virus. Vaccine 31:4305–4313

Steers NJ, Alving CR, Rao M (2008) Modulation of immunoproteasome subunits by liposomal lipid A. Vaccine 26:2849–2859

Steers NJ, Ratto-Kim S, De Souza MS et al (2012) HIV-1 envelope resistance to proteasomal cleavage: implications for vaccine induced immune responses. PLoS ONE 7:e42579

Stern LJ, Potolicchio I, Santambrogio L (2006) MHC class II compartment subtypes: structure and function. Curr Opin Immunol 18:64–69

Tyrrell DA, Heath TD, Colley CM et al (1976) New aspects of liposomes. Biochim Biophys Acta 457:259–302

Uemura K, Nicolotti RA, Six HR et al (1974) Antibody formation in response to liposomal model membranes sensitized with N-substituted phosphatidylethanolamine derivatives. Biochemistry 13:1572–1578

Vandepapeliere P (2018) Vaccine compositions comprising a saponin adjuvant, 4 Dec 2018

Verma JN, Wassef NM, Wirtz RA et al (1991) Phagocytosis of liposomes by macrophages: intracellular fate of liposomal malaria antigen. Biochim Biophys Acta 1066:229–238

Voltan AR, Alarcon KM, Fusco-Almeida AM et al (2017) Highlights in endocytosis of nanostructured systems. Curr Med Chem 24:1909–1929

Vu VP, Gifford GB, Chen F et al (2019) Immunoglobulin deposition on biomolecule corona determines complement opsonization efficiency of preclinical and clinical nanoparticles. Nat Nanotechnol 14:260–268

Wagner A, Vorauer-Uhl K (2011) Liposome technology for industrial purposes. J Drug Deliv 2011:591325

Waite DC, Jacobson EW, Ennis FA et al (2001) Three double-blind, randomized trials evaluating the safety and tolerance of different formulations of the saponin adjuvant QS-21. Vaccine 19:3957–3967

Wang N, Chen M, Wang T (2019) Liposomes used as a vaccine adjuvant-delivery system: from basics to clinical immunization. J Control Release 303:130–150

Wassef NM, Alving CR (1987) Complement-dependent phagocytosis of liposomes by macrophages. Methods Enzymol 149:124–134

Welsby I, Detienne S, N'kuli F et al (2016) Lysosome-dependent activation of human dendritic cells by the vaccine adjuvant QS-21. Front Immunol 7:663

White K, Krzych U, Gordon DM et al (1993) Induction of cytolytic and antibody responses using *Plasmodium falciparum* repeatless circumsporozoite protein encapsulated in liposomes. Vaccine 11:1341–1346

White WI, Cassatt DR, Madsen J et al (1995) Antibody and cytotoxic T-lymphocyte responses to a single liposome-associated peptide antigen. Vaccine 13:1111–1122

Wieczorek L, Krebs SJ, Kalyanaraman V et al (2015) Comparable antigenicity and immunogenicity of oligomeric forms of a novel, acute HIV-1 Subtype C gp145 envelope for use in preclinical and clinical vaccine research. J Virol 89:7478–7493

World Health Organization (2018) Malaria vaccine: WHO position paper, January 2016—recommendations. Vaccine 36:3576–3577

Young JF, Hockmeyer WT, Gross M et al (1985) Expression of *Plasmodium falciparum* circumsporozoite proteins in *Escherichia coli* for potential use in a human malaria vaccine. Science 228:958–962

Young JF, Ballou WR, Hockmeyer WT (1987) Developing a human malaria sporozoite vaccine. Microb Pathog 2:237–240

Zhou F, Watkins SC, Huang L (1994) Characterization and kinetics of MHC class I-restricted presentation of a soluble antigen delivered by liposomes. Immunobiology 190:35–52

Zurbriggen R (2003) Immunostimulating reconstituted influenza virosomes. Vaccine 21:921–924

# Polymeric Nanoparticle-Based Vaccine Adjuvants and Delivery Vehicles

**Elizabeth A. Grego, Alaric C. Siddoway, Metin Uz, Luman Liu, John C. Christiansen, Kathleen A. Ross, Sean M. Kelly, Surya K. Mallapragada, Michael J. Wannemuehler, and Balaji Narasimhan**

## Contents



---

Elizabeth A Grego, Alaric C Siddoway, Metin Uz—These authors contributed equally.

---

E. A. Grego · A. C. Siddoway · M. Uz · L. Liu · S. M. Kelly · S. K. Mallapragada · B. Narasimhan (✉)
Departments of Chemical and Biological Engineering, Iowa State University, Ames, IA 50011, USA
e-mail: nbalaji@iastate.edu

J. C. Christiansen · M. J. Wannemuehler
Departments of Veterinary Microbiology and Preventive Medicine, Iowa State University, Ames, IA 50011, USA

M. Uz · K. A. Ross · S. K. Mallapragada · M. J. Wannemuehler · B. Narasimhan
Departments of Nanovaccine Institute, Iowa State University, Ames, IA 50011, USA

Current Topics in Microbiology and Immunology (2021) 433: 29–76
https://doi.org/10.1007/82_2020_226

Published Online: 10 November 2020

**Abstract** As vaccine formulations have progressed from including live or attenuated strains of pathogenic components for enhanced safety, developing new adjuvants to more effectively generate adaptive immune responses has become necessary. In this context, polymeric nanoparticles have emerged as a promising platform with multiple advantages, including the dual capability of adjuvant and delivery vehicle, administration via multiple routes, induction of rapid and long-lived immunity, greater shelf-life at elevated temperatures, and enhanced patient compliance. This comprehensive review describes advances in nanoparticle-based vaccines (i.e., nanovaccines) with a particular focus on polymeric particles as adjuvants and delivery vehicles. Examples of the nanovaccine approach in respiratory infections, biodefense, and cancer are discussed.

## 1 Introduction

Vaccination plays a key role in preventive medicine by protecting individuals against harmful bacterial and viral diseases as well as for cancer immunotherapy. Recently, vaccine formulations have shifted away from whole bacteria or their lysates and inactivated viral particles towards highly purified recombinant protein antigens (Aoshi 2017; Leroux-Roels 2010). While these purified antigens allow for enhanced safety and targeting of the immune system towards specific epitopes, they are often poorly immunogenic compared to their live or attenuated counterparts (Aoshi 2017; Leroux-Roels 2010). Thus, adjuvants, or components that enhance the immune response, are an important consideration in modern vaccine design.

Adjuvants fulfill a wide variety of functions within vaccine formulations, with an overall goal to induce a potent immune response capable of providing long-term protection against future exposures (Montomoli et al. 2011). Adjuvants may act by directly stimulating immune cells via pattern recognition receptors or modulating the immune response to prioritize humoral or cell-mediated immunity (Montomoli et al. 2011; Coffman et al. 2010; Garlapati et al. 2009). Similarly, adjuvants may also be designed to overcome specific immune defects, such as immunosenescence in older adults, to improve vaccine efficacy (Leroux-Roels 2010; Montomoli et al. 2011). Another aspect of augmenting vaccine efficacy is through patient compliance. Adjuvants may not only provide immune stimulation but function as delivery vehicles capable of sustaining antigen release. The ability to enhance delivery and provide an antigenic depot allows for a reduction in doses, or the number of immunizations required, thereby enhancing patient compliance (Montomoli et al. 2011; Coffman et al. 2010). Furthermore, adjuvant vehicles may increase vaccine stability and shelf-life, allowing for a cost-effective vaccine to be deployed widely (Chen and Kristensen 2009). Thus, while adjuvants enable a wide variety of functions within vaccine formulations, multiple aspects must be considered when selecting the most appropriate adjuvant(s) for each vaccine application.

The focus of this chapter is on polymeric nanoparticle-based adjuvants, which provide multiple competitive advantages in the rational design of vaccines. By rationally selecting/designing polymers based on their physicochemical properties, and considering antigen and vaccine regimen, it is possible to modulate appropriate immune responses for specific diseases. We begin with a brief overview of the mechanisms of humoral- and cell-mediated immunity. In subsequent sections, the various types of polymeric nanoparticles that have been studied for vaccine use will be summarized and the advantages of natural and synthetic polymers in modulating immune response phenotypes will be described. Finally, examples of nanoparticle-based vaccines (or nanovaccines) against multiple diseases as well as advances in manufacture/scale-up of nanoparticle commercialization and regulatory considerations will be discussed.

## 2 Mechanisms of Immunity

### *2.1 Humoral Immunity*

Humoral immunity encompasses the functional capabilities of antibodies, complement cascade proteins, and antimicrobial peptides to eliminate extracellular and mucosal pathogens, signal innate immune cells, and enable immune protection. Antibodies have a wide range of functions including neutralizing virus and secreted toxins (McComb and Martchenko 2016; Klasse 2014), forming immune complexes to enhance complement activation, and binding to pathogens to promote cytolysis or phagocytosis by antigen-presenting cells (APCs) to activate $CD4^+$ and $CD8^+$ T cells (Wen et al. 2016).

Induction of antibody responses requires stimulation of B cells. B cells can initiate the production of T cell-independent antibodies in response to APC and T cell-derived cytokine stimulation or repetitive epitopes that cross-link B cell receptors (BCRs) (MacLennan et al. 2003). While antibodies produced this way can fix complement and are valuable in the early stages of an immune response, they have limited utility to meet the goals of vaccination due to their low affinity. These antibodies are not optimized for pathogen neutralization and the B cells that produce them are less likely to generate long-lived memory B cells and plasma cells.

Achieving protective and long-lived antibody production requires B cell enhancement by a subset of $CD4^+$ T cells called follicular helper T cells ($T_{fh}$). B cells in the germinal centers (GCs) compete to interact with follicular dendritic cells (FDCs) that present antigens and $T_{fh}$ cells. The cycling of B cell interactions with FDCs and $T_{fh}$ cells leads to antibody isotype class switching and affinity maturation of the BCRs. Those B cells with higher affinity preferentially repeat this cycle, leading to the production of high affinity, class-switched antibodies. $T_{fh}$ cell interactions also help signal B cells to leave GCs and differentiate into long-lived memory B cells and plasma cells.

### 2.2 *Cell-Mediated Immunity*

Cell-mediated immunity (CMI) involves activation of phagocytes [i.e., dendritic cells (DCs) and macrophages (MØs)], T helper (Th) cells, cytotoxic T lymphocytes (CTLs), and natural killer (NK) cells to eliminate pathogens. T cell activation requires interaction of the MHC: peptide complex on the surface of APCs with a complementary T cell receptor (TCR), signaling by other surface costimulatory receptor-ligand interactions, and cytokines provided by APCs. These cytokines are secreted as a consequence of APC interactions with the adjuvant during activation, allowing the adjuvant to indirectly determine the T cell phenotype (Wilson-Welder et al. 2009).

CMI is initiated by the internalization of antigen by DCs and MØs along with concomitant activation by the adjuvant; these APCs then migrate to the draining lymph nodes, present antigen to T cells in the context of MHC I and MHC II. Secretion of IL-4 and IL-2 promotes differentiation of naïve $CD4^+$ T cells towards a Th2 phenotype, which is subsequently involved in the humoral response (Saravia et al. 2019). DCs secreting IL-12 induces naïve $CD4^+$ T cell differentiation towards a $Th_1$ phenotype and naïve $CD8^+$ T cells towards a CTL phenotype (Kang et al. 2004). These $Th_1$ $CD4^+$ cells and CTLs then migrate to the site of vaccination/infection and interact with APCs capable of secreting IL-12, IFN-γ, TNF-α, and IL-2 (Vesely et al. 2011). Secreted IFN-γ activates MØs to enhance pathogen phagocytic ability, antimicrobial nitric oxide production, and antigen presentation to $CD4^+$ and $CD8^+$ T cells (Pennock et al. 2013). Additionally, differentiated CTLs have increased cytolytic capabilities through secretion of perforin and granzymes following cognate peptide-MHC I recognition on infected cells, which results in apoptosis (Annunziato et al. 2015).

## 3 Traditional Adjuvants

Currently, aluminum-based salts collectively referred to as 'alum', are one of the few adjuvants approved for human use. While alum has a long history of safe use in vaccines, it is far from a 'universal adjuvant' with a bias towards humoral immunity, a need for multiple doses, and incompatibility with many antigens (Gupta et al. 1995).

Mechanistically, alum promotes the recruitment of APCs, antigen uptake, and provides for a 'depot effect.' Antigen adsorbs onto alum and is slowly released into the surrounding microenvironment. These depots may persist for a significant amount of time and alum particles have been observed at the site of injection a year after administration (Gupta et al. 1995; Awate et al. 2013). Additionally, there is evidence that alum activates the innate immune system at the injection site. Research suggests alum binds to DC membrane lipids triggering the release of uric acid, which acts as a potent immunostimulant or damage-associated molecular pattern (DAMP) (Kool et al. 2008; Flach et al. 2011). Research has also found that alum may stimulate NLRP3 inflammasomes or induce apoptosis, which further activates immune responses (Quandt et al. 2015; Franchi and Núñez 2008).

Alum induces a strong $Th_2$ response, thus it is effective for disease platforms that require a humoral response for clearance. However, alum falls short in inducing $Th_1$ responses, a crucial response for clearing viral infections (Oleszycka et al. 2018; Igietseme et al. 2004). Additionally, alum is not a powerful immune stimulator; while not a significant handicap when working with killed or inactivated pathogens that provide a range of immunogenic components, this becomes a limitation when recombinant proteins are used for subunit vaccines. Typically, other immunogenic compounds must be co-administered in order to induce a protective response (McNeela and Mills 2001; Lavigne et al. 2004).

In addition to alum, monophosphoryl lipid A (MPLA) emulsions are also approved for human use (Reed et al. 2013). MPLA, a Toll-like receptor (TLR)-4 agonist, has been combined with a number of other TLR ligands to generate a wide range of adjuvant systems, many of which have undergone clinical testing (Table 1). These TLR ligands leverage specific pathways in the innate immune system rather than generalized inflammation. For example, the liposome adjuvant AS01 is a combination of MPLA and saponin QS-21. The resulting adjuvant system exhibits more balanced humoral and cellular responses than when either component is given alone (Didierlaurent et al. 2017).

While a complete review of all adjuvants lies beyond the scope of this chapter, it is sufficient to note that a select subset of adjuvants acts as TLR agonists. These agonists may be combined with other adjuvants/adjuvant systems or delivered via polymeric nanoparticles to further tailor the specific immune response elicited by the vaccine platform. Agonists to additional immune pathways, such as NOD-like receptors, C-type lectin receptors, folate receptors, and the STING pathway present more tools for fine-tuning the immune-enhancing capabilities of adjuvant systems (Garlapati et al. 2009; Chavez-Santoscoy et al. 2012; Narasimhan et al. 2016).

## 4 Polymer Nanoparticle Adjuvants

Polymer nanoparticles represent a promising adjuvant platform for a multitude of reasons (Fig. 1). One advantage of nanoparticle-based vaccines is that their biophysical and biochemical properties may be manipulated in order to enhance antigen uptake, processing, and presentation. Furthermore, rational selection of polymer chemistry (either natural or synthetic) may enhance antigen stability, influence release kinetics, and modulate the particular type of immune response. In this section, the overall physiochemical properties (including size, shape, charge, and hydrophobicity) of nanoparticle platforms will be described. In addition, detailed examples of the unique advantages of particle platforms (such as controlled release, co-encapsulation, and induction of CMI) will be summarized below.

**Table 1** List of adjuvants that have undergone clinical testing

| Adjuvant | Components | Formulation | Mechanism |
|---|---|---|---|
| Alhydrogel (aluminum hydroxide) | Alum | Aluminum adsorption | Depot effect, Recruitment |
| Alum (aluminum phosphate salts) | Alum | Aluminum adsorption | Depot effect, Recruitment |
| AS01 | MPL and QS-21 in liposomes<br>$AS01_B/AS01_E$ includes DOPC and cholesterol in phosphate buffered saline | Liposome | TLR-4, Recruitment |
| AS02 | MPL and QS-21 in liposomes | O/W emulsion | TLR-4 |
| AS03 | Oil-in-water emulsion with α-tocopherol | O/W emulsion | IRE1α |
| AS04 | MPL adsorbed onto aluminum hydroxide or aluminum phosphate | Aluminum adsorption | TLR-4 |
| AS15 | MPL, QS-21 and CpG 7909 | Liposome | TLR-4, TLR-9, Recruitment |
| AS25/AS50 | AS02 without QS-21 | O/W emulsion | TLR-4 |
| CAF01 | dimethyl dioctadecyl ammonium delivery vehicle | Liposome | Recruitment |
| CpG 7909 | | Soluble | TLR-9 |
| CRM197 | A modified diphtheria toxin | Soluble | Recruitment |
| dmLT | Proteins derived from *E. coli* enterotoxin | Soluble | TLR-4 |
| GLA-AF | Aqueous glucopyranosyl lipid adjuvant | O/W emulsion | TLR-4 |
| GLA-SE | Stable-oil-in-water-emulsion with glucopyranosyl lipid adjuvant | O/W emulsion | TLR-4 |
| GM-CSF | | Soluble | Recruitment |
| IC31 | Synthetic oligodeoxynuclotide and antibacterial peptide | Soluble | TLR-9 (Olafsdottir et al. 2009) |
| Type I interferon | | Soluble | Recruitment |
| IL-12 pDNA | | Soluble | Recruitment (Jalah et al. 2012) |
| IL-2 | | Soluble | Recruitment (Nunberg et al. 1989) |

(continued)

**Table 1** (continued)

| Adjuvant | Components | Formulation | Mechanism |
|---|---|---|---|
| ISCOMATRIX | Saponin and lipid complex | Immunostimulatory complex | Recruitment (Pearse and Drane 2005) |
| IsdB | Recombinant *S. aureus* protein | Soluble component | TLR-7 (Wu et al. 2014; Bagnoli et al. 2015) |
| KLH carrier protein | Immunogenic carrier protein | Soluble component | Recruitment |
| LT Adjuvant patch | Transcutaneous immunostimulatory patch | External microneedle | Recruitment (Mkrtichyan et al. 2011) |
| Matrix-M1 | Saponins | Soluble component | Antigen uptake and Recruitment (Magnusson et al. 2013) |
| MF59 | Squalene based emulsion | O/W emulsion | MyD88, ASC |
| Montanide ISA | Incomplete Freund's adjuvant in mannide monooleaste and mineral oil | W/O emulsion | Depot effect, Recruitment |
| MPL | | O/W emulsion | TLR-4 |
| MPL-SE | MPL plus glycerol, phosphatidylcholine and squalene | O/W emulsion | TLR-4 |
| poly-ICLC | Double stranded RNA complex | Soluble Component | TLR-3 (Saxena et al. 2019) |
| QS-21 (or OBI-821, OPT-821 or QS-DG all of which are equivalent derivatives or derived from different sources) | Saponins | Soluble component | Recruitment (Kensil and Kammer 2010) |
| rCTB | Cholera endotoxins | Soluble component | Recruitment (Isaka et al. 2004) |

## *4.1 Physiochemical Properties of Polymeric Nanoparticles*

### 4.1.1 Size

Particle size plays a large role in the distribution and cellular internalization of vaccine formulations. Particles in the 20–200 nm range can successfully enter the lymphatic system within a few hours of administration (Bachmann and Jennings

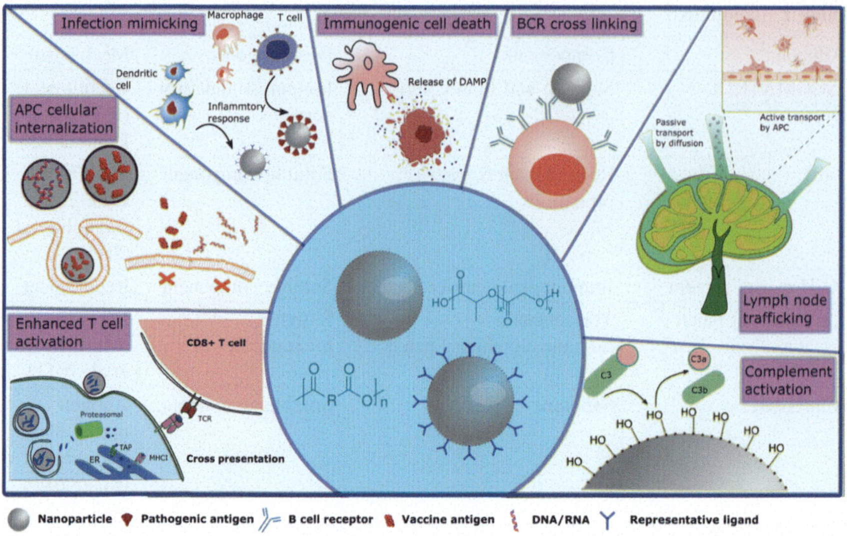


**Fig. 1** Schematic illustration of advantages provided by polymeric nanoparticle adjuvant systems for vaccination. Clockwise left to right: nanoparticle enhanced T cell activation, showing antigen cross-presentation; APC internalization of vaccine payload carried by nanoparticles versus free diffusion; nanoparticle-enabled infection mimicking inducing immune recognition and pro-inflammatory responses; immune activation by immunogenic cell death; nanoparticle-enabled B cell receptor (BCR) crosslinking; nanoparticle trafficking to lymph node, and complement activation by nanoadjuvants

2010). Particles with diameters of 200–500 nm cannot enter the lymphatic system directly but can be internalized by APCs and reach the lymphatic system within 24 hours (Bachmann and Jennings 2010). Particle size also affects the mechanism of internalization by APCs. Generally, sub-micron particles are internalized effectively by APCs and generate robust immune responses, while particles >10 μm cannot be internalized by APCs (Thomas et al. 2011; Ramirez et al. 2017). Particles with a diameter of 20–200 nm are internalized by endocytosis (Silva et al. 2013). In contrast, relatively larger particles (>0.5 μm) are internalized via micropinocytosis and phagocytosis by APCs (Silva et al. 2013).

### 4.1.2 Shape

While most studies focus on spherical particles, recent efforts have analyzed the effect of particle shape on immune responses. Particle shape influences cellular internalization, as it is suggested that rod-shaped particles may associate with and be internalized more readily by APCs than other particle geometries (He and Park 2016; Gratton et al. 2008). Experiments have shown improved cellular uptake of rod-shaped particles compared to spherical particles by DCs and MØs (Jindal 2017;

Sharma et al. 2010; Banerjee et al. 2016). Ex vivo studies have shown that rod-shaped nanoparticles can remain with the gastrointestinal tract for longer periods of time relative to spherical particles (Zhao et al. 2017). In vivo studies demonstrated that longer rod-shaped nanoparticles circulated in the blood longer than short-rod and spherical nanoparticles (Zhao et al. 2017).

### 4.1.3 Surface Charge

Both anionic and cationic particles are efficiently internalized by APCs, promoting induction of robust immunity (Thomas et al. 2011; Ramirez et al. 2017; Carrillo-Conde et al. 2012). However, intracellular tracking experiments show that internalized cationic nanoparticles escape from lysosomes in contrast to neutral and anionic particles that tend to localize within lysosomes (Yue et al. 2011). Therefore, particle charge can impact the ability of particles to influence induced immune response phenotype(s) by influencing the activation of APCs and targeting antigen to specific immune response pathways, which may mimic that of intracellular pathogens.

### 4.1.4 Hydrophobicity

Hydrophobicity has been suggested to act as a DAMP for activating innate immunity (Moyano et al. 2012). During host cell disruption, hydrophobic cellular materials are exposed to the external environment and trigger an innate immune response. It is hypothesized that increased surface hydrophobicity of nanoparticles may promote their interactions with hydrophobic components of cell surfaces to result in enhanced cellular uptake (Shima et al. 2013). In fact, hydrophobic polymeric particles synthesized with higher molecular weights tend to be phagocytosed more than their hydrophilic counterparts (Thomas et al. 2011; Shima et al. 2013; Goodman et al. 2014).

## *4.2 Unique Advantages of Nanoparticle Vaccines*

### 4.2.1 Controlled Release of Antigens

Polymeric particles can be designed to enable tunable release kinetics of antigens (Kumari et al. 2010). Depending on the synthesis method used, polymeric nanoparticles can be synthesized as nanospheres or nanocapsules. Nanospheres are comprised of a spherical matrix of polymer into which the antigen is physically entrapped. In nanocapsules, antigen is contained within a cavity surrounded by a polymer shell. Antigen can be loaded within the particles or onto their surface to manipulate release and subsequent immune effects. Unlike other delivery systems, polymeric nanoparticles are capable of encapsulating both hydrophilic and hydrophobic antigens (Kumari et al. 2010). Numerous factors contribute to antigen release rate from particles. For example, polymer hydrophobicity determines erosion kinetics, which in turn, determines antigen release kinetics, with hydrophobic surface eroding polymers exhibiting much slower release rates (Lopac et al. 2009). Combining physiochemical properties with chemistry allows for fine-tuned control. For example, short-rod nanoparticles exhibit a higher specific surface area than long-rod and spherical nanoparticles. When produced with a surface eroding polymer, the short-rod particles display shorter degradation time scales than their long-rod and spherical counterparts (Zhao et al. 2017).

### 4.2.2 Encapsulation of Co-adjuvants

The ability to encapsulate antigenic payloads into polymeric nanoparticles has been extensively shown to enhance the antibody response, allowing for both dose sparing strategies and reduced vaccination schedules (Huntimer et al. 2013a; Bershteyn et al. 2012). The physiochemical properties of polymeric nanosystems encompass both immunostimulatory adjuvant-like properties of the polymer and its chemical structure. The interplay between these two aspects leads to increased nanoparticle internalization by APCs, which enhances antigen uptake (Ulery et al. 2009). It has been demonstrated that some polymeric nanoparticle components are able to signal these pattern recognition receptors (PRRs) directly (Tamayo et al. 2010; Locatelli et al. 2017), or incorporate PRR stimulants (Phanse et al. 2017; Liu et al. 2018), to enhance antigenicity. By targeting various APC types with the right immunostimulatory signals through their physiochemical properties or by the inclusion of co-adjuvants (via co-administering small molecule adjuvants, co-encapsulation of adjuvant with antigen, or surface conjugation), it is possible to further enhance the magnitude of antibody responses, including generating neutralizing antibodies, as has been demonstrated for influenza and Ebola (Zacharias et al. 2018; Yang et al. 2017). The synergy in co-administration of multiple adjuvants in synthetic nanoparticle vaccines generates both increased and longer-lived humoral immune responses than when either adjuvant is used alone (Ulery et al. 2011; Stieneker et al. 1995; Wagner et al. 2019). It was demonstrated that co-administration of polymeric nanoparticles encapsulating TLR-4 and TLR-7 agonists induced neutralizing antibodies to hemagglutinin that were significantly higher in titer than those induced by either TLR agonist alone, and which persisted for over 1.5 years following immunization (Kasturi et al. 2011).

### 4.2.3 Cell-Mediated Immunity

Conventional live-attenuated vaccines or inactivated vaccines provide limited control over specific types of immune response as their biochemical properties are largely determined by the conserved nature of the pathogen itself. Additionally, vaccines using alum- or emulsion-based adjuvant strategies generally fall into generating either dominant humoral immunity or cause severe adverse reactions (Bhardwaj et al. 2020). To generate CMI, co-delivery of antigen mediated by both the MHC I and MHC II presentation pathways is key. Subunit antigens are generally poor inducers of $CD8^+$ T cells by themselves. Multiple polymeric nanoparticle formulations have demonstrated the ability to induce $CD8^+$ T cell responses (Huntimer et al. 2014; Zhang et al. 2011). This can be achieved through targeting of the nanoparticles to APCs, where the ability to activate these cells through both MHC II and MHC I presentation pathways can activate $CD4^+$ and $CD8^+$ T cells, respectively, and further allow $CD4^+$ T cells to secrete pro-inflammatory cytokines (e.g., IFN-$\gamma$) that enhance CTL responses, as well as by activating pathogen killing immune responses by innate immune cells. Mechanisms of nanoparticle adjuvanticity, including increased cellular uptake, enhanced cross-presentation, and immunogenic cell death (ICD) have been demonstrated to contribute to the induction of T cell activation (Lu et al. 2017; Urbanavicius et al. 2018). Studies have also shown induction of CMI by controlling nanoparticle polymer chemistry, hydrophobicity, size, charge, and pH responsiveness (Goodman et al. 2014; Gu et al. 2019; Rayahin et al. 2015; Luo et al. 2017). In addition, polymeric nanoparticles can incorporate co-adjuvants (as described above), which can further enhance $CD4^+$ and $CD8^+$ T cell responses (Zhang et al. 2011; De Titta et al. 2013; Hamdy et al. 2007).

## 5 Types of Polymeric Nanoparticle Adjuvants

Polymers are attractive materials for particle synthesis because of their biocompatibility, biodegradability, and low toxicity (Peres et al. 2017). Although inorganic particles such as gold, carbon, or silica have shown promising results as adjuvants, concerns over the potential risks of utilizing non-biodegradable materials remain (Pati et al. 2018; Tao and Gill 2015; Niikura et al. 2013; Kim et al. 2014). We next discuss both natural and synthetic polymers utilized for particle-based adjuvants, as well as hybrid polymer systems. This information is also summarized in Table 2.

### *5.1 Natural Polymers*

Natural polymers, obtained from renewable, sustainable, and natural resources, are considered as non-toxic, biocompatible, and biodegradable materials. Chitosan, alginate, dextran, hyaluronic acid, and inulin are the most common examples of natural polymers that are used as adjuvants in vaccine formulations (Torres et al. 2019; Bose et al. 2019; Guo et al. 2019; Nevagi et al. 2019a, b). Some of these natural polymers, such as β-glucans, are known to inherently target APCs and induce cellular and humoral immune responses, while some others, such as chitosan

**Table 2** Summary of polymer chemistries and characteristics

| Polymer | Structure | Advantages | Clinical Challenges | References |
|---|---|---|---|---|
| Chitosan | 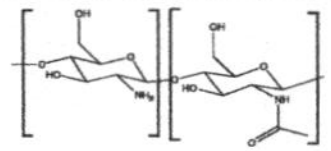 | Cationic, biodegradable, biocompatible, mucoadhesive immunomodulatory effects resulting in induction of both humoral and cellular immune responses | Insolubility and precipitation at physiological pH | Malik et al. (2018), Moran et al. (2018) |
| Alginate | 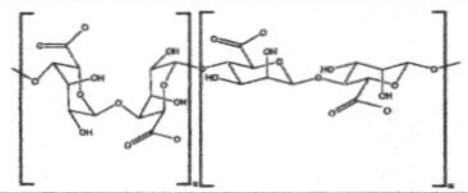 | Non-immunogenic and bioadhesive/mucoadhesive | Limited carrier properties dur to anionic nature | Ge et al. (2018) |
| Dextran | 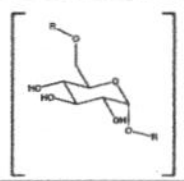 | Highly hydrophilic and water-soluble, induces strong humoral immunity | | Shen et al. (2013) |
| Cellulose | 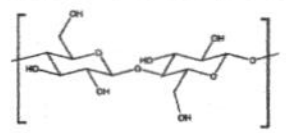 | Abundant and sustainable biopolymer | Could be toxic and induce immune reactions | Song et al. (2014) |
| Inulin | 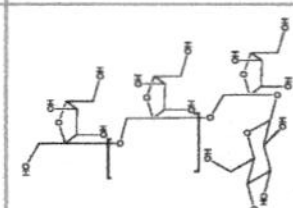 | Biocompatible biodegradable, and immunomodulatory properties | | Barclay et al. (2016) |
| Mannan | Mannan 1,4-polymaltose | Selective binding to mannose receptor, regulate DC trafficking and induce of both humoral and cellular responses | | Petrovsky and Cooper (2011) |
| Poly glutamic acid | 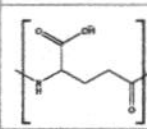 | Biodegradable, biocompatible, antigen-specific immune response | | Nevagi et al. (2019a, b) |
| Hyaluronic acid | 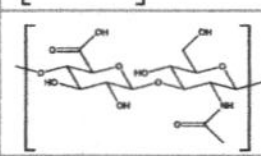 | Hydrophilic, biocompatible and biodegradable, transdermal immunization | | Kong et al. (2016) |

(continued)

**Table 2** (continued)

| Polymer | Structure | Advantages | Clinical Challenges | References |
|---|---|---|---|---|
| Poly(lactic-co-glycolic acid) | 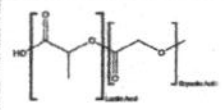 | Manipulation of hydrophobicity<br>Elicit IgG | Bulk erosion and strong acidity of degradation products | |
| Poly(lactic acid) | 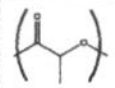 | Induced cytokine and chemokine secretion and increased $CD8^+$ T cell levels | Bulk erosion and strong acidity of degradation products | Thomas et al. (2011) |
| Poly(1,6-bis(*p*-carboxyphenoxy)-3,6-dioxaoctane) (CPTEG) | 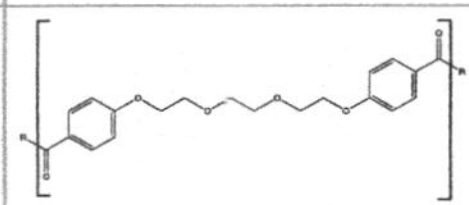 | Amphiphilic; leads to induction of strong CD4 and CD8 T cell responses | Relatively rapid degradation profile | Lopac et al. (2009) |
| Poly(1,6-bis(*p*-carboxyphenoxy) hexane) (CPH) | 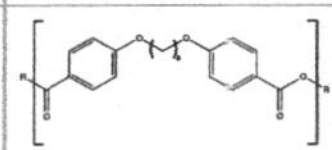 | Hydrophobic; presents strong danger signals to immune system | Relatively slow degradation profile | Lopac et al. (2009) |
| Poly(sebacic acid) (SA) | 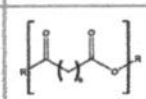 | Hydrophobic; internalized effectively by APCs | Acidic degradation products | |

The table shows the molecular structures of polymers discussed in this review and their advantages and clinical challenges

or alginate, are known for their mucoadhesive properties which may increase retention time (Petrovsky and Cooper 2011). Despite these advantages, natural polymers provide limited immune modulation compared to their synthetic counterparts (Guo et al. 2019). Nevertheless, the presence of specific chemical groups in their structure (i.e., amine or carboxyl groups) along with the anionic or cationic nature of these polymers enable different functionalities via alternative chemical and/or physical modifications. Thus, natural polymers can be modified by targeting ligands and receptors or loaded with antigens and other active agents to provide an adjuvanting effect and enhance immune response. Antigens and other active agents, such as nucleic acids, can be loaded via physical encapsulation, chemical adsorption, or electrostatic interactions to obtain natural polymer-based adjuvants and vaccine carriers in the forms of micro/nanoparticles, nanogels, nanofibers, and hydrogel capsules/nanoparticles. Numerous methods can be used to obtain these formulations, including solvent evaporation, emulsification-solvent diffusion, nanoprecipitation (Rao and Geckeler 2011), self-assembly, electrostatic complexation, ionic gelation, chemical crosslinking, template-assisted nanofabrication (Ferreira et al. 2013), electrospinning, (Jahantigh et al. 2014) or layer-by-layer assembly (Highton et al. 2015). These approaches are mainly used to provide high antigen loading and stability, which in turn improve delivery and immune response stimulation (Highton et al. 2015; Leleux and Roy 2013). Additionally, natural polymers can be used in combination with other synthetic or inorganic materials (i.e., synthetic polymers or inorganic particles) to develop hybrid systems with improved properties. These overall features of natural polymers make them attractive candidates as vaccine adjuvants (Han et al. 2018). We next describe some well-studied natural polymers that have been used in nanoparticle-based vaccine formulations.

### 5.1.1 Chitosan

Chitosan, obtained through deacetylation of chitin in shrimp and crustaceans, is a natural polymer commonly studied as a vaccine adjuvant. Highly acetylated chitosan is cationic in nature and is generally considered to be biodegradable, biocompatible and non-toxic. The main hurdle with chitosan is its insoluble nature at physiological pH. This may be addressed by additional deacetylation or chemical modification of amino moieties to obtain soluble derivatives of chitosan (i.e., trimethyl chitosan) (Moran et al. 2018).

Chitosan and its derivatives demonstrate inherent adjuvanting properties and immunomodulatory effects resulting in induction of both humoral and cellular immune responses (Nevagi et al. 2019a, b; Bose et al. 2019; Malik et al. 2018; Moran et al. 2018). Chitosan-based formulations have been reported to interact with various receptors on APCs (e.g., Dectin-1) and induce different signaling pathways involving NLRP3 inflammasome or cGAS-STING activation. These interactions trigger a cascade of different cellular events such as dendritic cell activation via type I interferon (IFN) resulting in secretion of chemokines (e.g., CXCL10/IP-10) and/or cytokines (e.g., TNF-α, IL-12, IL-4, and IL-10) and promoting Th1/Th2 immune response (Lampe et al. 2019). In addition, it has also been demonstrated that chitosan-based formulations can activate MØs (Malik et al. 2018; Moran et al. 2018), promote DC maturation, and enhance IgG and IgA antibody titers (Sui et al. 2010).

Chitosan and its derivatives can be used as nanoparticle adjuvants loaded with specific proteins, antigens, or active agents via ionic gelation. A study demonstrated that a chitosan nanoparticle-based Rift Valley Fever Virus vaccine formulation enhanced humoral and cellular immune responses by increasing cytokine secretion compared to its alum based counterparts (El-Sissi et al. 2020). Another study reported that a chitosan nanoparticle-based pneumococcal conjugate vaccine formulation provided enhanced immune responses by increasing IgG1, IgG2a, IgG2b, and IgG3 antibody titers compared to liposome nanoparticle-based formulations (Haryono et al. 2017). In classical approaches, chitosan nanoparticle formulations encapsulating proteins acted as vaccine adjuvants by boosting cytokine secretion, promoting Th1/Th2 responses, increasing natural killer cells activity and humoral/cellular immune response (Wen et al. 2011).

Chitosan based nanovaccine formulations have demonstrated mucoadhesive and adjuvanting properties (Xia et al. 2015). Considering this, a study developed chitosan and trimethyl chitosan adjuvants for intranasal H5N1 influenza vaccines which resulted in significant antibody responses. They also reported that the formulation with trimethyl chitosan demonstrated no clinical signs post-challenge (Mann et al. 2014). With this perspective, another study intranasally administered early secreted antigenic target 6-kDa protein (ESAT-6) encapsulated in trimethyl chitosan (TMC) nanoparticles which induced IgG against *Mycobacterium tuberculosis* (Amini et al. 2017). As a new approach, a study used mast cell activator C48/80 in combination with chitosan nanoparticles as an alternative vaccine adjuvant for Hepatitis B via the nasal route (Bento et al. 2019). Finally, encapsulation of antigens in chitosan-based vaccine formulations also provides stability and antigenicity in the gastrointestinal tract following oral immunization (Farhadian et al. 2015).

### 5.1.2 Alginate

Alginate is a biodegradable, biocompatible, muchoadhesive, natural polymer, and is commonly used as an adjuvant in vaccine formulations. The nanoparticle form of hydrophilic alginate provides efficient antigen release and improved immunogenicity particularly via nasal and oral administration (Guo et al. 2019; Nevagi et al. 2019a; Sarei et al. 2013). It has been reported that alginate incorporated influenza vaccine formulations induced cytokine production, increased antibody titers, activated DCs, and enhanced cellular immunity (Ge et al. 2018; Dehghan et al. 2018). Negatively charged alginate was also used to coat chitosan nanoparticles loaded with inactivated influenza virus to modulate immunostimulatory properties. Intranasal administration of this vaccine formulation resulted in an increased Th1 immune response (Mosafer et al. 2019). Similar strategies have also been used to develop alginate-based vaccines for hepatitis B and A providing improved humoral and cellular immune responses (AbdelAllah et al. 2016; AbdelAllah et al. 2020).

### 5.1.3 Dextran

Dextran and its derivatives [e.g., diethylaminoethyl (DEAE)-dextran or acetylated dextran (Ac-DEX)] are hydrophilic and water-soluble natural polymers, demonstrating adjuvant properties and providing robust immune responses (Moreno-Mendieta et al. 2017; Bachelder et al. 2017). It was reported that dextran nanoparticles incorporating ovalbumin and lipopolysaccharide promoted mannose receptor-dependent APC stimulation, inducing both cellular and humoral immune responses in mice (Shen et al. 2013; De Geest et al. 2012). In a recent study, ovalbumin encapsulated Ac-DEX microparticles demonstrated controlled release of active cargo accompanied by efficient antibody and cytokine production (Chen et al. 2018a). The same group also developed Ac-DEX, matrix protein 2, and 3′3′-cyclic GMP-AMP (cGAMP)-based vaccine formulation and reported enhanced humoral and cellular responses resulting in protection against lethal influenza (Chen et al. 2018b). In a follow-up study, the same group encapsulated STING agonist cGAMP and soluble Toll-like receptor 7/8 (TLR7/8) agonist resiquimod (R848) into Ac-DEX microparticles that enabled enhanced cellular and humoral responses (Collier et al. 2018). Similar approaches with Ac-DEX and cGAMP were also used by the same group to further support their findings (Junkins et al. 2018). Another group also used Ac-DEX to encapsulate TLR-7 agonist imiquimod and deliver it to intracellular TLR-7 receptors resulting in an increase of IL-1β, IL-6, and TNF-α (Bachelder et al. 2010). Chemical conjugation of dextran with type B CpG DNA, a TLR-9 agonist, was also conducted resulting in enhanced APC internalization, improved CpG accumulation in lymph node, and improved $CD8^+$ T cell responses in mice (Zhang et al. 2017a).

### 5.1.4 Hyaluronic Acid

Hyaluronic acid, a hydrophilic, biocompatible, and biodegradable natural mucopolysaccharide (Ekici et al. 2011) can be used as a adjuvant stabilizer and delivery vehicle to be applied via the transdermal route due to its involvement in skin extracellular matrix and ability to penetrate the skin due to its small size (Nashchekina and Raydan 2018; Bussio et al. 2019). Hyaluronic acid also improves solubility of adjuvants or antigens, such as MPLA and alum adjuvanted hepatitis B vaccine, enhancing both cellular and humoral immune responses (Moon et al. 2015).

### 5.1.5 Inulin

Inulin is another biocompatible and biodegradable natural biopolymer demonstrating vaccine adjuvant properties (Barclay et al. 2016; Cooper et al. 2013; Cooper and Petrovsky 2011; Cooper et al. 2015; Kumar and Tummala 2013). Acetylated inulin (Ace-IN) is most commonly used in vaccine formulations and has been shown to trigger cytokine secretion, DC maturation, antibody titer production (Rajput et al. 2018). An Ace-IN-based pathogen-mimicking vaccine delivery system was reported to improve antigen delivery to APCs and simultaneously activate TLR-4 on APCs (Kumar et al. 2017; Kumar et al. 2016).

### 5.1.6 DNA

Nucleic acids or DNA-based nanomaterials have also been used as vaccine adjuvants through functionalization with short immunogenic sequences. For example, it was reported that CpG-containing dendrimer like-DNA showed shape and size-dependent immunostimulatory activity (Mohri et al. 2015; Mohri et al. 2014). Differently, self-assembled single-stranded DNA sequences (e.g., immune nanoflowers) as well as DNA hydrogels have also been developed to provide cytokine secretion and antigen delivery (Zhang et al. 2015; Ishii-Mizuno et al. 2017). Recently, DNA origami-based approaches have been developed to provide immune stimulation based on different origami geometries and shapes (Hong et al. 2017; Bila et al. 2019; Bastings et al. 2018). It was reported that compact shapes possessing low-aspect ratios are efficiently internalized while non-compact shapes remained on the cell surface, demonstrating the ability to affect DC uptake and activation by rational design (Bastings et al. 2018).

### 5.1.7 Other Natural Polymers

Cellulose-based materials have also been used in vaccine formulations as adjuvants for proteins, antigens, or DNA in the form of particles, nanowires, or nanofibers to enhance the immune response via increasing the secretion of pro-inflammatory cytokines (Song et al. 2014; Bin et al. 2018; Catalan et al. 2010; Čolić et al. 2015; Tomić et al. 2016; Vartiainen et al. 2011; Pereira et al. 2013).

Antigen-containing natural, biodegradable, biocompatible, and non-toxic anionic poly(glutamic acid), γ-PGA, nanoparticles have also shown significant antigen-specific immune responses by activation of $CD8^+$ T cells (Uto et al. 2013) and human monocyte-derived DCs (Broos et al. 2010). The γ-PGA particles are generally modified with L-PAE and are potential antigen carriers with excellent adjuvant properties via TLR-4, activation enhancing both humoral and cellular immunity (Uto et al. 2015; Ikeda et al. 2018).

Other biopolymers demonstrating adjuvant effects in vaccine formulations are mannan (Apostolopoulos et al. 2013), lentinan (Zhang et al. 2017b), zymosan

(Ainai et al. 2006), pullulan (Singh et al. 2017), and carrageenan (Zhang et al. 2010) which all provided immunomodulation via of different signaling pathways that enhanced humoral and cellular immune responses.

## 5.2 *Synthetic Polymers*

Synthetic polymers serve as a promising vaccine adjuvant materials because of their biocompatibility, biodegradability, and low cost. The following sections will discuss polymeric platforms such as polyesters, polyanhydrides, and synthetic micelles.

### 5.2.1 Polyesters

Numerous polyesters have been studied for nanoparticle vaccines; however, poly (lactic-co-glycolic acid) (PLGA) is the most well-studied because it is biodegradable and exhibits high biocompatibility (Silva et al. 2016; Danhier et al. 2012). Due to a favorable safety profile, PLGA has received FDA approval for various biomedical applications, including sutures and drug and vaccine delivery (Silva et al. 2016; Danhier et al. 2012). The release rate of the cargo and hydrophobicity of PLGA particles can be directly manipulated by varying the ratio of lactic acid to glycolic acid monomers (Allahyari and Mohit 2016). PLGA particles release encapsulated material through a bulk erosion mechanism, resulting in burst-like release profiles. PLGA particles with higher amounts of glycolic acid result in a larger and more rapid burst release of antigen (Thomas et al. 2011; Allahyari and Mohit 2016).

When considering the effects of PLGA particles on the immune system, PLGA particles have shown success in presenting antigen to stimulate both humoral and cellular immunity (Nicolete et al. 2011). Multiple studies have shown successful internalization of PLGA particles by DCs and MØs (Danhier et al. 2012; Semete et al. 2010; Liu et al. 2015). Following internalization by DCs, PLGA particles release antigen for presentation through the MHC I pathway to elicit $CD8^+$ T cell responses (Liu et al. 2015). Studies have shown PLGA enhanced numbers of $CD4^+$ and $CD8^+$ T cells in the spleen when compared to administration of antigen alone (Liu et al. 2015). These findings were also supported by other studies that showed activation of $CD4^+$ and $CD8^+$ T cells by antigen-encapsulated in PLGA nanoparticles (Demento et al. 2012).

PLGA particles also result in enhanced cytokine secretion profiles from DCs. Hepatitis B antigen-encapsulated into PLGA particles showed a significant increase in IL-2 and IFN-$\gamma$ when compared to antigen alone (Thomas et al. 2011). Antigen-encapsulated PLGA particles also induced higher amounts of IL-1$\beta$, IL-6, and IFN-$\alpha$ (Demento et al. 2012). Mice immunized with *Bordetella pertussis* encapsulated in PLGA particles showed a significant enhancement of IFN-γ and IL-17, suggesting a mixed $Th_1$ and $Th_{17}$ immune response (Li et al. 2016).

With respect to humoral immunity, PLGA nanoparticles have shown efficacy at eliciting significant levels of systemic IgG and mucosal IgA. Mice injected with OVA-loaded PLGA particles reached a peak concentration of antigen-specific IgG antibodies at week 6 post-immunization that was maintained for 13 weeks (Demento et al. 2012). Orally administered PLGA particles co-loaded with OVA and the immunostimulant, MPLA, maintained an 8-fold increase in IgG titers for 5 weeks when compared to OVA alone (Sarti et al. 2011). PLGA nanoparticles loaded with OVA and CpG elicited enhanced production of OVA-specific IgG2a antibodies when compared to OVA and CpG alone (Joshi et al. 2013). In addition, enhanced titers of mucosal IgA were observed upon administration of hepatitis B antigen-encapsulated in PLGA particles (Thomas et al. 2011).

### 5.2.2 Polyanhydrides

Polyanhydrides are another class of biodegradable polymers with high biocompatibility (Huntimer et al. 2013b). This class of polymers has been approved for human use in medical controlled release products such as Gliadel® and Septacin (Roy et al. 2016; Wafa et al. 2019). Polyanhydrides degrade by surface erosion into non-toxic and easily metabolized carboxylic by-products (Lopac et al. 2009). During erosion, polyanhydride particles degrade only at the surface and exclude water from the bulk material, in contrast to the bulk erosion exhibited by polyesters (Silva et al. 2013). This surface erosion phenomenon contributes to tunable and sustained release profiles of encapsulated payloads. Polyanhydride hydrophobicity (and antigen release kinetics) can be directly manipulated by varying the ratio of its copolymer constituents. For example, copolymers rich in 1,6-bis(*p*-carboxyphenoxy)hexane (CPH) are largely hydrophobic and release antigen over a period of months (Kipper et al. 2002). In contrast, copolymers rich in 1,8-bis(*p*-carboxyphenoxy)-3,6-dioxaoctane (CPTEG) or sebacic anhydride (SA) degrade more quickly (ca. days-weeks).

Polyanhydride particles are also advantageous in terms of their enhanced antigen stability (Determan et al. 2006; Carrillo-Conde et al. 2010; Haughney et al. 2013; Ross et al. 2014; Vela Ramirez et al. 2016; McGill et al. 2018), sustained antigen release, and activation of B cells and T cells (inherent adjuvanticity), all of which lead to induction of robust humoral and cell-mediated immunity (Vela Ramirez et al. 2016; Huntimer et al. 2013a, 2014; Ramirez et al. 2014; Vela Ramirez 2015). Another significant benefit of polyanhydride nanovaccines is their superior thermal stability at room temperature for extended periods of time (Shen et al. 2001; Petersen et al. 2011; Wagner-Muñiz et al. 2018).

Polyanhydride particles are internalized by and activate APCs (Ulery et al. 2009). CPH, CPTEG, and SA-based particles have been shown to be internalized by bone marrow-derived DCs and MØs (Wafa et al. 2019; Torres et al. 2011).

Generally, hydrophobic particles are internalized more readily by APCs (Ramirez et al. 2017; Petersen et al. 2011; Phanse et al. 2016). DCs or MØs exposed to polyanhydride particles upregulated MHC I and II, costimulatory molecules such as CD40, CD80, and CD86, and cytokine secretion (e.g., IL-6, IL-12p70, IL-1β, TNF-α) (Wafa et al. 2019; Torres et al. 2011; Phanse et al. 2016; Wafa et al. 2017).

Polyanhydride nanovaccines have been shown to induce robust humoral and cell-mediated immunity. For example, it was shown that a single dose of a polyanhydride nanovaccine with a 64-fold less dose of antigen as compared to soluble antigen induced similar antibody titers (Huntimer et al. 2013a). Studies have demonstrated the successful formation of GCs by a single administration of CPTEG:CPH nanoparticles (Vela Ramirez et al. 2016). Polyanhydride nanovaccines have been shown to induce sustained levels of both antigen-specific IgG and IgA antibodies (Zacharias et al. 2018; Wafa et al. 2017; Haughney et al. 2014). Finally, a single dose of polyanhydride nanovaccine was able to induce neutralizing antibodies against influenza virus in mice and pigs (Ross et al. 2015; Dhakal et al. 2017a, b; 2019). Polyanhydride nanovaccine formulations have been shown to promote both $CD4^+$ and $CD8^+$ T cell responses (Zacharias et al. 2018; Huntimer et al. 2013a, 2014). Following intranasal immunization, these studies demonstrated the induction of tissue-resident memory phenotypes that are believed to be important in conferring heterosubtypic protective immunity against influenza virus infection (Zacharias et al. 2018).

Another advantage of polyanhydride particles is their easily functionalized surfaces. Studies have shown that attachment of sugars to polyanhydride nanoparticles assist in targeting CLRs on APCs (Carrillo-Conde et al. 2012). Functionalization of polyanhydride particles with di-mannose significantly enhanced particle internalization by DCs (Ramirez et al. 2014; Phanse et al. 2013). In addition, these di-mannose functionalized particles upregulated MHC II, CD86, and CD40 expression on DCs (Carrillo-Conde et al. 2011). Further studies determined that di-mannose functionalized polyanhydride particles also upregulated CD206, a CLR, on bone marrow-derived DCs (Ramirez et al. 2014). Functionalized nanoparticles were also shown to elevate IL-6 and TNF-α secretion from DCs (Carrillo-Conde et al. 2011).

### 5.2.3 Micelles

Synthetic polymers may also be used to produce micelles. Micelles contain a hydrophobic core (for holding payloads) and a hydrophilic shell/corona. Similar to polymeric particles, micelles are capable of delivering antigen while maintaining adjuvant-like properties (Trent et al. 2015).

Micelles can be synthesized from di-block or multi-block copolymers. Typical polymers used to form the hydrophilic shell include polyethylene glycol (PEG) and polyethylene oxide (PEO). Polymers used to form the hydrophobic core include poly(L-amino acids), polyesters, and phospholipids. Phospholipids can be produced through both synthetic and natural means. When exposed to block-selective solvents, block copolymers undergo spontaneous self-assembly to form micelles (Abetz 2005). The specific shape of the micelle is determined by the type of block copolymer used and interactions between the micelle core and block-selective solvent (Abetz 2005).

In vivo studies have shown that administration of antigen-loaded micelles elicits an immune response similar to antigen administered with a TLR-2 ligand (Trent et al. 2015). In addition, it was shown that the adjuvanticity of the micelle-peptide formulation was directly related to the delivery of the peptide as opposed to induction of cytokines or costimulatory molecule upregulation. Micelles, which are typically less than 100 nm, are successful at delivering antigen to DCs (Trent et al. 2015). Due to their relatively small size, micelles are capable of both interacting with DCs at the site of delivery and transiting lymphatic vessels directly to the lymph nodes (Facciolà et al. 2019).

A family of amphiphilic pentablock copolymers with controlled architectures possessing different cationic end blocks has been reported as potential vaccine carriers for subunit vaccines (Adams et al. 2014, 2015, 2019). Particularly, a pentablock copolymer, composed of a temperature-responsive Pluronic F127 middle block and two pH-responsive poly(diethyl aminoethyl methacrylate) end blocks, demonstrated temperature and pH-responsive micellization and gelation which forms a depot for controlled delivery of proteins and genes (Adams et al. 2014, 2015, 2019). The amphiphilic central block, Pluronic F127, is an effective vaccine adjuvant promoting cellular entry and contributes to gene delivery while the cationic outer blocks spontaneously condense negatively charged DNA via electrostatic interactions for sustained combinational therapy. This polymer was also further modified with mannose to enhance interactions with CD206, improving its adjuvanting properties for DNA-based vaccine delivery (Adams et al. 2014). In addition, this pentablock polymer demonstrated in vivo persistence at the injection site for over 50 days providing sustained protein release and five-fold enhancement in the antigen-specific antibody titer compared to soluble protein alone (Adams et al. 2015). Recently, the high biocompatibility of these injectable Pluronic-PDEAEM micelle adjuvants was demonstrated, further enhancing their potential as components of next-generation vaccines (Adams et al. 2019).

## *5.3 Hybrid Polymer Adjuvants*

Hybrid systems composed of different classes of materials, such as natural/synthetic polymers and lipids have been developed as adjuvants for vaccine formulations. Using PEG as surfactant is one of the most commonly applied approaches enhancing the stability and stealth properties of the systems. To increase mucoadhesive and mucopenetrating properties of vaccines via nasal administration, a study modified the surface of chitosan-based nano-emulsions with PEG (Di Cola et al. 2019). Another common approach is to synergistically use carbohydrates to coat synthetic polymer surfaces to enhance cellular uptake. For example, mannosylated poly(beta-amino esters) (PBAEs), free of inflammatory co-adjuvants, provided enhanced gene delivery resulting in improved APC activation and immune response (Jones et al. 2015). In another example, alginate was used to form cationic nanogels disulfide cross-linked with polyethylenimine (PEI) for delivery of OVA. It was reported that the OVA-containing nanogels significantly improved the induction of tumor-specific $CD8^+$ T cells and antigen-specific antibody production (Li et al. 2013).

The cationic nature of chitosan and its derivatives can be used to coat synthetic polymer particles to increase antigen loading efficiency and/or adjuvanticity (Wusiman et al. 2019). For example, PLGA nanoparticles, loaded with Heptatis B surface antigen (HBsAg) and coated with trimethyl chitosan (TMC), were reported to provide efficient absorption and increase antibody titers via nasal route (Pawar et al. 2010). Similarly, chitosan-coated PLA particles enhanced antigen adsorption capacity and macrophage uptake resulting upregulation of MHC I and MHC II and secretion of pro-inflammatory cytokines (Chen et al. 2014). In comparison to a commercially available vaccine, another study demonstrated that coating poly-ε-caprolactone (PCL) with chitosan resulted in elevated IgG responses against HBsAg while avoiding the induction of IgE (Jesus et al. 2018). Furthermore, the nanoparticle formulation induced CMI against HBsAg leading to IL-17 and IFN-γ secretion. Along these lines, B and T cell epitope-containing peptides were coupled with PGA and mixed with TMC to produce self-adjuvanting nanovaccines (Nevagi et al. 2019b).

Hybrid polymer systems can also combine disparate polymer systems to achieve synergistic immune responses for a wide range of pathogens and delivery routes. In a recent study, an increased humoral immune response upon oral delivery was achieved by developing a hybrid delivery system in which self-assembled lipopeptides nanoparticles were coated with alginate and TMC layers via electrostatic interactions (Bartlett et al. 2020). Similarly, a cationic liposome and natural hyaluronic acid-based hybrid nanoparticle system was shown to provide improved stability and antigen release characteristics that significantly enhanced the serum IgG response to *Y. pestis* F1-V following intranasal administration (Fan et al. 2015). In another study, liposome-PEG-PEI complex-based adjuvant demonstrated enhanced uptake, expression of surface markers, induction of pro-inflammatory cytokines, and antigen presentation (Chen et al. 2012). It was also shown that intranasal immunization with chitosan-coated, lipid-polymer hybrid nanoparticles enhanced mucosal immune responses via induction of both humoral and cell mediated immunity (Rose et al. 2018). Liposomes, modified with pH-sensitive polymers, could also serve as a delivery system for the induction of antigen-specific $CD8^+$ T cells. For example, inclusion of cationic lipids into polymer-modified liposomes promoted costimulatory molecule expression and secretion of IL-12 and TNF-α from DCs and induced antigen-specific secretion of IFN-γ from splenocytes (Yoshizaki et al. 2017).

# 6 Polymeric Nanovaccines Against Diseases

Polymeric nanovaccines have demonstrated great potential as vaccine adjuvants against both viral and bacterial pathogens. In this section, we include examples of the use of polymeric nanovaccines against infectious diseases, biodefense pathogens, and cancer. These examples demonstrate how nanovaccines can be tailored to be effective countermeasures against each of these significant health challenges.

## *6.1 Infectious Diseases*

Infectious diseases such as influenza, coronavirus disease, and infections caused by respiratory syncytial virus (RSV) continue to pose great threats to public health. These pathogens are often highly contagious or can be transmitted from natural reservoirs, which makes disease eradication extremely difficult. Their high mutation rate and variety of subtypes render traditional inactivated or live-attenuated vaccines unlikely to provide long-term protection to susceptible populations that vary by age, immune heritage, and underlying health conditions. The COVID-19 pandemic has underlined the need for development of novel vaccination strategies capable of inducing long-lasting, heterosubtypic (i.e., cross-protective) immunity. In this regard, nanovaccines demonstrate unique advantages because of their flexibility with respect to administration routes, ability to co-deliver multiple payloads, enhanced thermal stability, and induction of rapid and long-lived immunity in a single dose, as described below.

### 6.1.1 Influenza

Influenza remains a serious source of morbidity and mortality worldwide despite decades of research towards developing a vaccine for both seasonal and pandemic strains. Typical 'flu shots' can successfully build strain-specific humoral immunity, but fail to initiate robust CMI necessary for clearing heterologous viral infections (Ho et al. 2011; Muruganandah et al. 2018). The performance of seasonal vaccinations is affected by the high mutation rate of influenza strains, with tetravalent vaccines rarely reaching 50% efficacy, compounded with poor immune responses in at-risk populations, immunosenescence in older adults and immunocompromised individuals (Demicheli et al. 2018; Ohmit et al. 2013; Della Bella et al. 2007). In addition to immunological shortcomings, current vaccines are also limited by manufacturing capacity, antigenic changes (i.e., glycosylation) and specificity associated with 70-year old egg-based technologies, and dependence on a cold chain for both storage and distribution (Wu et al. 2017; Zost et al. 2017).

Polymeric nanovaccines present a powerful alternative to current influenza vaccinations. Acting in a dual capacity as adjuvants and delivery systems for diverse payloads (influenza virus antigens such as hemagglutinin, nucleoprotein, neuraminidase, etc. and additional co-adjuvants), polymeric nanovaccines can deliver multiple formulations of vaccines tailored to either specific age groups, or particular patient needs (Ross et al. 2019). These nanovaccines benefit from rapid and large-scale production, delivering a thermostable product that is cost-effective to produce, stockpile, and deploy (Kelly et al. 2020). To this end, polymers composed of polysaccharides, polyesters, and polyanhydrides have been used to formulate nanoparticles to deliver influenza virus antigens (Ross et al. 2015; Dhakal et al. 2017a, b; Renu et al. 2020). These versatile nanovaccines have been successfully delivered intranasally, intramuscularly, and subcutaneously. For respiratory pathogens, such as influenza virus, intranasal immunizations with nanovaccines have been shown to generate local tissue-resident T cell immunity, which leads to protection against heterologous challenge. This is a distinct advantage over current influenza vaccines (Lau et al. 2012), setting the stage for a "universal flu vaccine". Intranasal vaccinations mimic both the natural site and route of infection, leveraging the natural immune response to protect the body from future infections (Ross et al. 2015).

### 6.1.2 Respiratory Syncytial Virus

Human respiratory syncytial virus (RSV) infects epithelial cells of the lungs to cause potentially severe upper and lower respiratory infections in newborns, young children, immunosuppressed, and the aged with no approved vaccines available (Gilbert et al. 2018; Tang et al. 2019; Swanson et al. 2020). RSV has multiple surface proteins (attachment glycoprotein (G) and fusion (F) glycoprotein) which are necessary for viral cell attachment and fusion to epithelial cells for successful infection. These proteins are viable vaccine targets due to their necessity for infection and multiple epitopes for neutralization (Swanson et al. 2020). However, these proteins are poorly immunogenic, complicating the formulation of a successful vaccine (McGill et al. 2017). Although the FDA has approved a monoclonal antibody, palivizumab, solely as a preventative treatment reserved for high-risk infants, the use of palivizumab can promote the escape of mutant RSV strains resistant to neutralization (Gilbert et al. 2018).

Multiple studies have shown robust immune response in mice and rats promoted by F protein encapsulated polyanhydride nanoparticles (Gilbert et al. 2018; Stephens et al. 2019, 2020). Other studies have shown the success of co-encapsulating G and F proteins into polyanhydride nanoparticles and administered to neonatal calves to induce protection from bovine respiratory syncytial virus (BRSV) (McGill et al. 2017, 2018). Calves serve as an excellent animal model because BRSV infection in calves mimics RSV infection in humans (McGill et al. 2017). All of these studies attest to the strong potential of RSV antigen-based nanovaccines as a viable approach to protect against RSV infections.

### 6.1.3 Coronavirus

Coronaviruses are responsible for multiple respiratory infections, including severe acute respiratory syndrome (SARS-CoV), Middle East Respiratory Syndrome (MERS-CoV), and most recently, SARS-CoV-2, which causes coronavirus disease (COVID-19) (Coleman and Frieman 2014; Rothan and Byrareddy 2020). These infections are responsible for severe complications in older adults and the immune-compromised. Coronaviruses are characterized by surface spike proteins responsible for viral entry (Coleman et al. 2014). The spike protein also serves as a promising target for vaccines. Due to the contagious nature of SARS-CoV-2 transmission between humans, there is an immediate need to develop a safe and effective vaccine.

Recent studies evaluating intranasal delivery of SARS-CoV S DNA encapsulated within polyethylimine nanoparticles demonstrated induction of antigen-specific humoral and cellular immune responses in mice (Shim et al. 2010). The DNA-containing nanoparticles induced enhanced production of IgA as well as CD80, CD86, and MHC II expression on DCs that likely contributed to an increase in polyfunctional $CD4^+$ and $CD8^+$ T cells. These studies present a promising avenue to combat coronaviruses, including the development of a COVID-19 nanovaccine.

### 6.1.4 Pneumonia

*Streptococcus pneumoniae* is the biological agent that causes pneumococcal pneumonia. Pneumococcal infections are high-risk to both young and aged populations, and among the deadliest in children under the age of five. The capsular polysaccharide of this organism is highly variable, accounting for 98 serotypes, making it difficult to develop a 'universal' vaccine (Masomian et al. 2020). Although there currently are three licensed vaccines for pneumococcal pneumonia, namely PPV23, PCV10, and PCV13, their introduction has inadvertently increased disease prevalence caused by strains expressing non-vaccine serotypes. This has led to renewed interest in developing a vaccine that more broadly protects individuals (Masomian et al. 2020). Recent approaches include the pneumococcal surface protein A (PspA) for which there are only three families representing six clades (Piao et al. 2014). Strains expressing family 1 of 2 PspA represent approximately 96% of clinical isolates. For this reason, vaccines incorporating PspA proteins may overcome the challenges posed by the use of capsular vaccines. In this regard, polyanhydride nanoparticles encapsulating PspA induced high titer antigen-specific IgG and demonstrated the ability to protect mice from pneumococcal challenge (Wagner-Muñiz et al. 2018). Compared to PspA alone, this vaccine formulation provided a high level of protection using a 25-fold dose reduction and was shelf-stable for at least 60 days at room temperature. Chitosan nanoparticles have also been used to intranasally deliver DNA encoding pneumococcal PsaA, demonstrating the ability to induce mucosal and systemic antibody responses to

PsaA, as well as reduced bacterial burden in the nasopharynx of challenged mice (Xu et al. 2011). Chitosan-based nanoparticles containing PspA have been shown to induce antigen-specific IgA and IgG as well as CMI; following challenge, mice were protected against otitis media and against intraperitoneal challenge with both serotypes 3 and 14 (Xu et al. 2015).

## 6.2 *Biodefense Pathogens*

Developing new vaccines to counteract biodefense pathogens is a high priority. Of these pathogens, tier 1 select agents pose the greatest risk to military and civilian populations. These include numerous toxins, viral and bacterial organisms, including *Yersinia pestis*, *Bacillus anthracis*, and Ebola virus. Polymeric nanoparticles represent an attractive vaccine platform against these deadly biological agents. Subunit vaccines are ideal biodefense vaccines due to concerns of adverse reactions or incomplete inactivation of dangerous biological organisms. However, subunit antigens are generally poorly immunogenic. Hence, the physiochemical characteristics of nanoparticles enable the design of vaccine formulations that can adjuvant subunit antigens and induce protective immune responses through activating multiple arms of adaptive immunity. The following is a brief overview of nanovaccine development against multiple biodefense pathogens.

### 6.2.1 Pneumonic Plague

*Yersinia pestis* is a tier 1 select agent that causes plague, which can manifest as pneumonic, bubonic, and septicemic forms (Riedel 2005). F1-V, a fusion protein of the F1 capsular and the V antigen (part of the type III secretion appendage) of *Y. pestis*, has been investigated as a subunit vaccine candidate. Our laboratory has previously reported on polyanhydride nanoparticles encapsulating F1-V that induced high avidity IgG1 responses in immunized mice, which persisted over 280 days post-immunization and protected vaccinated mice against lethal challenge (Ulery et al. 2011). We have also reported on a combination adjuvant-based nanovaccine containing F1-V encapsulated into polyanhydride nanoparticles co-adjuvanted with cyclic di-GMP (cdG), which also protected mice against lethal challenge with *Y. pestis* CO92 in as early as 14 days post-immunization (Wagner et al. 2019).

### 6.2.2 Anthrax

*B. anthracis* is a tier 1 select agent that is the causative agent of anthrax. This organism can form durable spores that can infect humans through multiple routes, including inhalation, subcutaneous, and gastrointestinal (WHO 2008). Studies using polyanhydride nanoparticles encapsulating the *B. anthracis* protective antigen (PA) demonstrated the ability to maintain the structural stability of encapsulated PA (Petersen et al. 2012). The conformational stability of PA is important for generating toxin-neutralizing antibody responses. This stability was maintained for at least four months following storage at temperatures up to 40 °C (Petersen et al. 2012). Additionally, PLGA nanoparticles encapsulating domain IV of PA have been shown to induce a mixed $Th_1/Th_2$ immune response as evidenced by the in vitro secretion of IL-1 and IFN-$\gamma$ (Manish et al. 2013).

### 6.2.3 Ebola Virus

Ebola virus is a tier 1 select agent that causes Ebola virus disease. This disease is zoonotic, and can be transmitted from person-to-person via direct contact, infected bodily fluids, or contaminated fomites (Jacob et al. 2020). Polymeric nanoparticles have been used as a DNA vaccine adjuvant and delivery platform against Ebola (Yang et al. 2017). Ebola DNA (EboDNA) was coated onto cationic PLGA-poly-L-lysine/poly-$\gamma$-glutamic acid nanoparticles and administered to mice using a microneedle patch or through intramuscular injection. The microneedles dissolve within five minutes, allowing the particles to disseminate and induce immune responses, resulting in an enhanced magnitude of neutralizing IgG antibodies to Ebola virus.

### 6.2.4 Brucella abortus

*Brucella abortus* is a select agent that causes brucellosis. Its primary host species are cattle, and it is capable of infecting humans as an incidental host, usually through direct contact or consumption of animal products such as non-pasteurized dairy products (De Figueiredo et al. 2015). PLGA nanoparticles have been investigated as a vaccine platform against *B. abortus* (Singh et al. 2015). Nanoparticles encapsulating rL7/L12 ribosomal protein demonstrated the ability to induce high IgG1 antibody responses as well as secretion of IFN-$\gamma$ from lymphocytes following ex vivo stimulation. This resulted in protection from bacterial challenge, as demonstrated by decreased bacterial load in the spleen.

## *6.3 Cancer*

Both therapeutic and prophylactic cancer vaccines need immune-stimulatory adjuvants that can break tolerance and induce humoral and cell-mediated immune responses to tumor-associated antigens (TAAs). To overcome the challenges of immune dysfunction associated with the tumor microenvironment, various small molecule adjuvants or nano-adjuvanted therapies have been tested in clinical and preclinical settings. The first examples of FDA-approved or late-stage clinical nanotherapeutic endeavors were liposome-based platforms. The applications of nano-enabled therapies have been pioneered by liposome-based drug delivery systems, as discussed by excellent reviews (Shi et al. 2017; Wicki et al. 2015; Nam et al. 2019). Here, we focus on translational examples of adjuvant molecules and polymeric nanoparticulate systems with the potential for cancer vaccines.

### 6.3.1 Breast Cancer

Breast cancer is the most common malignancy worldwide. There are many recently concluded or ongoing clinical trials for the application of granulocyte-monocyte colony-stimulating factor (GM-CSF), one of the most widely used cytokine-based adjuvants, in vaccines for breast cancer patients. Many of these efforts are focused on the HER2 antigen that is expressed in about 15–30% of invasive breast cancer. In a series of clinical trials with HER2 derived peptide, the GM-CST adjuvanted vaccine was used as an adjunct therapy to prevent tumor recurrence after the completion of standard-of-care therapies. These vaccines, incorporating HER2 peptides E75, GP2, AE37, augmented $CD8^+$ or $CD4^+$ T lymphocyte responses and demonstrated acceptable safety and tolerability (Clifton et al. 2015). All three vaccines were advanced into phase II or III clinical investigations and revealed increased but not significant disease-free 5-year survival compared to control patients (Mittendorf et al. 2014, 2016a, b).

Nanogel delivery vehicles formulated with adjuvants have been designed to aid HER2 vaccine immunotherapy (Neamtu et al. 2017). The nanogels release polypeptide payload upon swelling in water and induce improved T cell activation via enhanced cellular internalization and cytosolic antigen presentation (Gu et al. 1998). In phase I clinical trial, self-assembled amphiphilic cholesteryl pullulan nanogels encapsulating truncated HER2 antigen were formulated with GM-CSF or OK-432 adjuvant. It was demonstrated that five out of nine patients with various types of solid tumors were characterized with HER2-specific $CD8^+$/$CD4^+$ T cell responses (Kitano et al. 2006). A similar nanogel was used in another phase I clinical trial characterizing humoral immunity. In 14 of the 15 patients studied, HER2-specific IgG antibodies were measurable in patient sera (Kageyama et al. 2008).

More nanovaccine formulations are being tested for breast cancer in preclinical animal models (Allahverdiyev et al. 2018). One example describes the activation of DCs and the induction of tumor-specific $CD8^+$ T cells following administration of CpG-coated PLGA nanoparticles loaded with tumor antigen (Kokate et al. 2016). This functional co-delivery system (i.e., tumor antigen and TLR agonist) was suggested to specifically interact with DCs in a bacteriomimetic manner (Kokate et al. 2015). The multifunctional immunostimulatory capability of these nanoformulations distinguishes them from conventional vaccination.

### 6.3.2 Lung Cancer

Liposome-based lung cancer vaccines have been advanced into clinical study. Tecemotide (L-BLP25) is a MUC1 glycoprotein immunotherapy liposomal vaccine combined with MPLA that is capable of inducing antigen-specific T cell responses (Mehta et al. 2012). The vaccine formulation was proven to induce a dominant $Th_1$ response and CTL specific to MUC1. Multiple phase II clinical studies were carried out with systemic analysis of disease prognosis from late-stage unresectable non-small lung cancer (NSCLC) patients (Butts et al. 2005, 2010, 2011; Wu et al. 2011). Although phase III trials concluded similar performance after L-BLP25 administration as placebo, in contrast to the mild adverse reactions and improved median survival observed in phase II trials (Butts et al. 2014). Thus, more liposomal vaccine studies are worth pursuing depending upon the stage of cancer relative to the treatment regimen employed. In a recent example using a murine model, hyaluronic acid nanoparticles loaded with microRNA-125b were used to reprogram M2 phenotype MØs in a NSCLC model. These particles were previously shown to target $CD44^+$ MØs and modified to enable negatively charged nucleotide encapsulation (Jain et al. 2015). Nanoparticle accumulation in lung tissue was confirmed. Repolarization of tumor-associated MØs to M1 phenotype was demonstrated by altered surface biomarker expression.

### 6.3.3 Pancreatic Cancer

Pancreatic cancer (PC) is characterized by a dense stromal barrier and dysregulated immune cells, causing resistance to traditional therapies and immunomodulation. Immunoadjuvant nanoparticulate systems have been applied to counter such barriers and sensitize PC for immunotherapy. Lu et al. prepared a lipid bilayer encapsulating mesoporous silica nanoparticles that co-deliver an immunogenic cell death (ICD) inducing drug and indoleamine 2,3-dioxygenase (IDO) pathway inhibitor to induce anti-tumor immunity (Lu et al. 2017). The nano-delivery formulation was proposed to prolong ICD potency and simultaneously reverse immunosuppression mediated by the IDO signaling pathway. Improved survival of tumor-bearing mice was demonstrated along with increased numbers of tumor-infiltrating immune cells. Shen et al. reported on the design of lipid-protamine nanoparticles loaded with DNA encoding trap proteins for IL-10 and CXCL12, two major proteins that contribute to the tumor immunosuppression (Shen et al. 2018). The perivascular delivery of nanoparticles induced local expression of trap proteins and reversal of immunosuppression, also indicated by increased numbers of tumor-infiltrating DCs, $CD8^+$ T cells, and NK cells. This nanoformulation showed improved survival in a KPC murine model, further underlining the potential of nano-therapies to treat PC.

### 6.3.4 Brain Cancer

Glioblastoma multiforme (GBM) is the most common and deadly glioma with an incidence rate of 4–5 per 100,000 persons and a dismal 5% five-year survival rate (Ostrom et al. 2014). While surgical removal of the tumor can delay disease progression, residual cancerous cells often result in tumor reoccurrence. Attempts to address these cells with systemically administered chemotherapeutics have been limited by bioavailability across the blood–brain barrier (BBB), short circulation half-life, and toxic off-target effects, leading to brief and irregular dosage at the tumor site (Wait et al. 2015). The Stupp method combines resection, external beam radiation, and systemically administered temozolomine (TMZ), and has become the standard of care for GBM. However, mean survival of patients was about 14 weeks as reported in 2005 (Stupp et al. 2005).

To overcome these limitations, a local sustained release platform was developed using polyanhydride (CPP:SA) wafers loaded with carmustine (Tamargo et al. 1993). This product, known as the GLIADEL® wafer, is placed by the surgeon directly into the tumor bed post-resection. The surface eroding properties of the wafer protect carmustine from degradation and produce a sustained release, while direct placement in the intracerebral cavity circumvents the BBB and avoids systemic, off-target toxicity (Tamargo et al. 1993; Brem et al. 1994). These advantages have borne out in clinical trials where carmustine-loaded wafers increase mean survival of both recurrent and newly diagnosed high-grade glioma patients, either as a monotherapy or in combination therapy (Wait et al. 2015).

While the GLIADEL® wafer is a great example of a polymer-based drug delivery system, it has some limitations. First, the wafer must be placed directly in the brain, limiting its application to immediate use after surgery. To address this, nanoparticles capable of crossing the BBB and delivering carmustine to target locations are necessary (Wadajkar et al. 2017). Second, the GLIADEL® wafer does not enhance the immune response to GBM or contribute to long-term protective immunity. Several nanoformulations target DCs to enhance antigen presentation and CTL activation, leading to successful treatment in glioma animal models. Poly (β-L-malic acid) based nanoscale immunoconjugates (NICs) covalently linked to anti-PD-1 and anti-CLTA-4 monoclonal antibodies have been evaluated for their ability to deliver checkpoint therapy across the BBB. In rats treated with the NICs, there was an increase of CTLs, NK cells, and M1 MØs in the tumor accompanied by a decrease in $T_{REG}$ cells, and an increase in the length of survival (Galstyan et al. 2019). These studies demonstrated the feasibility of delivering nanoscale immunodrugs across the BBB to treat brain tumors and other neurodegenerative diseases.

## 7 Manufacturing, Scale-up, and Regulatory Considerations

While nanoparticle adjuvants clearly demonstrate immunological and delivery benefits, it is important to consider all aspects of vaccine development for translation from research and development to commercialization. A crucial aspect of vaccine development is the ability to maintain stability and viability over long periods of time and/or variable temperatures. Exposure of vaccines to elevated temperatures may result in denaturation of proteins and loss of potency (Kumru et al. 2014; Lloyd and Cheyne 2017). Similarly, exposure to freezing temperatures can be just as destructive, inducing agglomeration of adjuvants and precipitation (Kumru et al. 2014; Lloyd and Cheyne 2017; Hanson et al. 2017). Thus, the vaccine cold chain, a network of maintaining refrigerated temperatures (2–8 °C) through manufacturing, storage, and distribution, is required for most vaccines.

The vaccine cold chain, while largely successful, is not without its challenges. Power outages, failing, and outdated equipment, as well as transportation delays, may result in exposure to increased temperatures (Chen and Kristensen 2009). Traditionally, vaccine vial monitors indicate storage exceeded elevated temperatures, which consequently leads to vaccine wastage (Chen and Kristensen 2009; Kumru et al. 2014; Lloyd and Cheyne 2017). However, freeze exposure is often less obvious and undetected, despite an estimate of at least 75% of vaccine shipments being exposed to freezing temperatures (Chen and Kristensen 2009). In addition, maintaining the vaccine cold chain is increasingly expensive, accounting for up to 80% of the cost of each dose (Chen and Kristensen 2009; Lee et al. 2017). As such, countries in need of basic vaccines often lack the resources or infrastructure to be able to successfully maintain the cold chain (Chen and Kristensen 2009; Lee et al. 2017).

Recently, researchers have turned their focus to designing thermostable vaccines through the use of particle adjuvants and delivery vehicles. Although liquid formulations have thus far been preferred due to ease of manufacturing and packaging, dry formulations via spray drying are an increasingly appealing option to enhance vaccine thermostability. As an example, Saboo et al. demonstrated that a spray-dried virus-like particle (VLP) vaccine maintained stability and immunogenicity after storage at 37 °C for up to 14 months (Saboo et al. 2016). Similarly, adsorption to or encapsulation of vaccines within polymer matrices may enhance vaccine thermostability. Recently, a vaccine for Ebola virus coated onto PLGA particles was found to maintain stability for six weeks at 37 °C (Yang et al. 2017). Additionally, PLGA particles have been demonstrated to preserve the stability of encapsulated inactivated poliovirus (Tzeng et al. 2016). Furthermore, polyanhydride nanoadjuvants have demonstrated the ability to maintain the stability of multiple subunit vaccine antigens (Carrillo-Conde et al. 2010; Haughney et al. 2013; Ross et al. 2014). In contrast to alum, Petersen et al. showed encapsulation into polyanhydride nanoparticles preserved PA from *B. anthracis* over a wide range of temperatures (−20 to 40 °C) for up to four months (Petersen et al. 2012). Thus, nanoadjuvants are a promising platform to enhance the thermostability of vaccines and eliminate the vaccine cold chain.

Despite their improved thermostability, commercial manufacturing of nanovaccines includes many challenges. Current vaccine production facilities are highly specific to a single, unique product and new manufacturing facilities would be needed (Hosangadi et al. 2020). Although this may require significant capital upfront, many nanovaccines offer a "plug-and-play" platform approach which lends flexibility to the manufacturing process of different vaccine formulations (Hosangadi et al. 2020). In addition, current dry vaccines utilize lyophilization, a lengthy, high energy, and expensive process which requires reconstitution before administration (Amorij et al. 2008; Clements and Wesselingh 2005; McAdams et al. 2012; Plotkin et al. 2017). High throughput, cost-effective production of nanovaccines could be achieved via spray drying (McAdams et al. 2012; Jin and Tsao 2013). However, unlike the food and pharmaceutical industries, current good manufacturing practices (cGMP) are not yet established for vaccine production (McAdams et al. 2012; Jin and Tsao 2013). Thus, further exploration is needed to translate nanovaccines from laboratory benchtop to commercial scale.

Commonly employed techniques for nanoparticle synthesis include emulsion-based methods, nanoprecipitation, salting out, microfluidics, and spray drying (Paliwal et al. 2014; Sosnik and Seremeta 2015). Benchtop methods such as single and double emulsion methods, as well as nanoprecipitation, have traditionally been widely employed in the design of polymeric nanoparticles encapsulating a wide variety of payloads, however, there are issues with its scalability, including batch-to-batch variability and high cost due to the requirement of large solvent volumes, and extensive drying following synthesis to ensure loss of solvent, all of which can become costly on the material and processing end and require large containers to accommodate such volumes (Martínez Rivas et al. 2017; Truong-Dinh Tran et al. 2016). These limitations, as well as investor skepticism about the feasibility of scaling up emulsion-based methods and nanoprecipitation, have made it difficult to scale-up for commercial use.

As mentioned above, spray drying is an alternative method for nanoparticle synthesis that is gaining interest for large-scale production. This method is used to synthesize polymeric micro- and nanoparticles encapsulating proteins in a variety of polymer chemistries, including the widely used PLGA (Brenza et al. 2015; Arpagaus 2019; Sharma et al. 2015), but also for emulsions and micelles. Spray drying is a method of forming particles through atomization of a liquid containing polymer and protein, either co-dissolved in solution, emulsified, or as a solid/oil suspension. The main advantages of spray drying over conventional polymeric nanoparticle syntheses are that it is rapid, continuous, low energy, easily scalable, has high ease of operation, is versatile to multiple formulations, single-step, and does not require any purification or further processing (Sosnik and Seremeta 2015; Arpagaus et al. 2018). In addition, multiple studies have documented high yields using this process, capable of achieving yields well above 50% (Schmid et al. 2011; Draheim et al. 2015; Li et al. 2010; Shi et al. 2020).

Currently, the ability to spray dry nanoparticles on a commercial scale is limited. For benchtop nanoparticle synthesis, the Büchi Nano Spray Dryer B-90 HP is capable of spray drying particles $\geq 200$ nm by atomizing particles via ultrasonic vibrating nozzle and collected via an electrical precipitator (Wisniewski 2015). However, translation of this technology to pilot and eventually commercial scale is challenging as industrial-scale spray dryers of this nature do not currently exist. That being said, microparticle spray drying is a much more developed field of study with technology established to commercial scale. Several options for laboratory benchtop models exist, including the Büchi B-290 Mini Spray Dryer which atomizes particles via a two-fluid nozzle and collects particles with a cyclone. In addition, Büchi Corporation published a report on the scalability from the B-290 Micro Spray Dryer to the pilot plant spray dryer Niro MOBILE (Arpagaus and Schwartzbach 2008). A previous study demonstrated the scalability of influenza virus-derived hemagglutinin microparticle (5–10 μm) formulation from the B-290 to the pilot-scale spray dryer BLD-1 (0.5–30 μm) (Zhu et al. 2014). Therefore, the avenue exists to scale the production of smaller sized microparticles, which is required by the United States Food and Drug Administration (FDA), and there is a great need for research into developing similar pathways for spray dryers that can produce nanoparticles on a pilot-scale and beyond.

It is also important to partner with regulatory agencies to more carefully understand scale-up and manufacturing of nanovaccines. Whereas much effort has been placed toward the research and development of nanovaccines, regulatory guidance and processes for nanoparticles are still evolving. Indeed, any changes in vaccine manufacturing induce new regulatory requirements. Often, even small changes require re-validation of clinical trials before licensing the vaccine product (Plotkin et al. 2017; Hua et al. 2018; Zehrung et al. 2017), demonstrating a need for standardized rapid, in vitro assays that can ensure nanoparticle safety and efficacy (Hua et al. 2018). Similarly, new administration devices will also need to meet regulatory agency requirements and may need different pathways for approval (Zehrung et al. 2017).

Although commercialization of nanovaccines presents some challenges, it is important to realize the full impact nanovaccines may have on preserving health worldwide. In recent years, pandemics have highlighted a need for preparedness. As mentioned, the "plug-and-play" manufacturing of nanovaccines offers flexibility, allowing facilities to be easily modified to meet urgent needs (Hosangadi et al. 2020). Needle-free administration would enable the ability to self-vaccinate without highly trained healthcare professionals, providing a cost-effective method for rapid vaccination (Hosangadi et al. 2020; Jin and Tsao 2013; Zehrung et al. 2017; Lee et al. 2015). As an example, the ability to self-administer an influenza vaccine has been reported to increase the intent to vaccinate by approximately 20%, and reduce societal costs by \$2.6 billion US dollars (Lee et al. 2015). Furthermore, the enhanced thermostability of nanovaccines would allow for large, inexpensive stockpiles ready for deployment in critical public health crises, such as pandemics (Hosangadi et al. 2020; Amorij et al. 2008; McAdams et al. 2012; Lee et al. 2015).

# 8 Conclusions and Perspectives

Nanovaccines represent a promising platform for many of today's vaccines and for future vaccines. In particular, polymeric adjuvants offer multiple advantages in designing novel vaccines to help counter emerging and re-emerging pathogens (including ones that can cause pandemics), non-communicative diseases such as cancer, and biodefense pathogens in both humans and animals. As described in this review, polymeric nanovaccines can reduce the dose and number of immunizations for many vaccines, provide rapid and long-lived immunity in a single dose, activate humoral and cell-mediated immunity, provide for room temperature storage for long periods of time, and enhanced patient compliance. Many of these advantages also translate very effectively to the design and development of novel vaccines for livestock and companion animal vaccines. With the advent of the data science revolution, new approaches can be developed for rational design of novel nanovaccine formulations using artificial intelligence and data analytics. Challenges that still need to be overcome include translation from lab to clinic, scale-up/manufacturing/costs, and regulatory approvals. Altogether, polymeric nanovaccines represent a platform technology with great promise to enhance public health and pandemic preparedness.

**Acknowledgements** The authors acknowledge partial financial support from the National Institutes of Health (R01 AI111466-01, R01 AI127565-01A1, R01 AI141196-01, U01 CA213862-01, and R01 HD095880-01A1) and the Nanovaccine Institute. B.N. acknowledges the Vlasta Klima Balloun Faculty Chair and S.K.M. is grateful to the Carol Vohs Johnson Chair.

**Disclosures** Balaji Narasimhan and Michael Wannemuehler are co-founders of ImmunoNanoMed Inc., a start-up with business interests in the development of nano-based vaccines against infectious diseases. Narasimhan also has a financial interest in Degimflex LLC. Surya Mallapragada is a co-founder of Degimflex LLC., a start-up with business interests in the development of flexible degradable electronic films for biomedical applications. She also has a financial interest in ImmunoNanoMed Inc.

# References

AbdelAllah NH, Abdeltawab NF, Boseila AA, Amin MA (2016) Chitosan and sodium alginate combinations are alternative, efficient, and safe natural adjuvant systems for hepatitis B vaccine in mouse model. Evidence-Based Complement Altern Med 2016:7659684

AbdelAllah NH, Gaber Y, Rashed ME et al (2020) Alginate-coated chitosan nanoparticles act as effective adjuvant for hepatitis A vaccine in mice. Int J Biol Macromol 152:904–912

Abetz V (2005) Block copolymers I. Springer-Verlag, Berlin Heidelberg

Adams JR, Goswami M, Pohl NLB, Mallapragada SK (2014) Synthesis and functionalization of virus-mimicking cationic block copolymers with pathogen-associated carbohydrates as potential vaccine adjuvants. RSC Adv 4:15655–15663

Adams JR, Haughney SL, Mallapragada SK (2015) Effective polymer adjuvants for sustained delivery of protein subunit vaccines. Acta Biomater 14:104–114

Adams JR, Senapati S, Haughney SL et al (2019) Safety and biocompatibility of injectable vaccine adjuvants composed of thermogelling block copolymer gels. J Biomed Mater Res—Part A 107:1754–1762

Ainai A, Ichinohe T, Tamura S et al (2006) Zymosan enhances the mucosal adjuvant activity of poly(I:C) in a nasal influenza vaccine. Antivir Ther 55:52–55

Allahverdiyev A, Tari G, Bagirova M, Abamo ES (2018) Current approaches in development of immunotherapeutic vaccines for breast cancer. J Breast Cancer 21:343–353

Allahyari M, Mohit E (2016) Peptide/protein vaccine delivery system based on PLGA particles. Hum Vaccines Immunother 12:806–828

Amini Y, Tebianian M, Mosavari N et al (2017) Development of an effective delivery system for intranasal immunization against Mycobacterium tuberculosis ESAT-6 antigen. Artif Cells, Nanomed Biotechnol 45:291–296

Amorij JP, Huckriede A, Wilschut J et al (2008) Development of stable influenza vaccine powder formulations: challenges and possibilities. Pharm Res 25:1256–1273

Annunziato F, Romagnani C, Romagnani S (2015) The 3 major types of innate and adaptive cell-mediated effector immunity. J Allergy Clin Immunol 135:626–635

Aoshi T (2017) Modes of action for mucosal vaccine adjuvants. Viral Immunol 30:463–470

Apostolopoulos V, Thalhammer T, Tzakos AG, Stojanovska L (2013) Targeting antigens to dendritic cell receptors for vaccine development. J Drug Deliv 2013:1–22

Arpagaus C (2019) PLA/PLGA nanoparticles prepared by nano spray drying. J Pharm Investig 49:405–426

Arpagaus C, Schwartzbach H (2008) Scale-up from bench-top research to laboratory production. Büchi Inf Bull 8

Arpagaus C, Collenberg A, Rütti D et al (2018) Nano spray drying for encapsulation of pharmaceuticals. Int J Pharm 546:194–214

Awate S, Babiuk LA, Mutwiri G (2013) Mechanisms of action of adjuvants. Front Immunol 4:1–10

Bachelder EM, Beaudette TT, Broaders KE et al (2010) In vitro analysis of acetalated dextran microparticles as a potent delivery platform for vaccine adjuvants. Mol Pharm 7:826–835

Bachelder EM, Pino EN, Ainslie KM (2017) Acetalated dextran: a tunable and acid-labile biopolymer with facile synthesis and a range of applications. Chem Rev 117:1915–1926

Bachmann MF, Jennings GT (2010) Vaccine delivery: a matter of size, geometry, kinetics and molecular patterns. Nat Rev Immunol 10:787–796

Bagnoli F, Fontana MR, Soldaini E et al (2015) Vaccine composition formulated with a novel TLR7-dependent adjuvant induces high and broad protection against *Staphylococcus aureus*. Proc Natl Acad Sci U S A 112:3680–3685

Banerjee A, Qi J, Gogoi R et al (2016) Role of nanoparticle size, shape and surface chemistry in oral drug delivery. J Control Release 238:176–185

Barclay TG, Rajapaksha H, Thilagam A et al (2016) Physical characterization and in silico modeling of inulin polymer conformation during vaccine adjuvant particle formation. Carbohydr Polym 143:108–115

Bartlett S, Eichenberger RM, Nevagi RJ et al (2020) Lipopeptide-based oral vaccine against hookworm infection. J Infect Dis 221:934–942

Bastings MMC, Anastassacos FM, Ponnuswamy N et al (2018) Modulation of the cellular uptake of DNA origami through control over mass and shape. Nano Lett 18:3557–3564

Bento D, Jesus S, Lebre F et al (2019) Chitosan plus compound 48/80: formulation and preliminary evaluation as a hepatitis B vaccine adjuvant. Pharmaceutics 11:72

Bershteyn A, Hanson MC, Crespo MP et al (2012) Robust IgG responses to nanograms of antigen using a biomimetic lipid-coated particle vaccine. J Control Release 157:354–365

Bhardwaj P, Bhatia E, Sharma S et al (2020) Advancements in prophylactic and therapeutic nanovaccines. Acta Biomater 108:1–21

Bila H, Kurisinkal EE, Bastings MMC (2019) Engineering a stable future for DNA-origami as a biomaterial. Biomater Sci 7:532–541

Bin Lee H, Yoon SY, Singh B et al (2018) Oral immunization of FMDV vaccine using pH-sensitive and mucoadhesive thiolated cellulose acetate phthalate microparticles. Tissue Eng Regen Med 15:1–11

Bose RJC, Kim M, Chang JH et al (2019) Biodegradable polymers for modern vaccine development. J Ind Eng Chem 77:12–24

Brem H, Tamargo RJ, Olivi A et al (1994) Biodegradable polymers for controlled delivery of chemotherapy with and without radiation therapy in the monkey brain. J Neurosurg 80:283–290

Brenza TM, Petersen LK, Zhang Y et al (2015) Pulmonary biodistribution and cellular uptake of intranasally administered monodisperse particles. Pharm Res 32:1368–1382

Broos S, Lundberg K, Akagi T et al (2010) Immunomodulatory nanoparticles as adjuvants and allergen-delivery system to human dendritic cells: implications for specific immunotherapy. Vaccine 28:5075–5085

Bussio JI, Molina-Perea C, González-Aramundiz JV (2019) Hyaluronic acid nanocapsules as a platform for needle-free vaccination. Pharmaceutics 11:246

Butts C, Murray N, Maksymiuk A et al (2005) Randomized phase IIB trial of BLP25 liposome vaccine in stage IIIB and IV non-small-cell lung cancer. J Clin Oncol 23:6674–6681

Butts C, Murray RN, Smith C et al (2010) A multicenter open-label study to assess the safety of a new formulation of BLP25 liposome vaccine in patients with unresectable stage III non-small-cell lung cancer. Clin Lung Cancer 11:391–395

Butts C, Maksymiuk A, Goss G et al (2011) Updated survival analysis in patients with stage IIIB or IV non-small-cell lung cancer receiving BLP25 liposome vaccine (L-BLP25): Phase IIB randomized, multicenter, open-label trial. J Cancer Res Clin Oncol 137:1337–1342

Butts C, Socinski MA, Mitchell PL et al (2014) Tecemotide (L-BLP25) versus placebo after chemoradiotherapy for stage III non-small-cell lung cancer (START): a randomised, double-blind, phase 3 trial. Lancet Oncol 15:59–68

Carrillo-Conde B, Schiltz E, Yu J et al (2010) Encapsulation into amphiphilic polyanhydride microparticles stabilizes *Yersinia pestis* antigens. Acta Biomater 6:3110–3119

Carrillo-Conde B, Song EH, Chavez-Santoscoy A et al (2011) Mannose-functionalized "Pathogen-like" polyanhydride nanoparticles target C-type lectin receptors on dendritic cells. Mol Pharm 8:1877–1886

Carrillo-Conde BR, Roychoudhury R, Chavez-Santoscoy AV et al (2012) High-throughput synthesis of carbohydrates and functionalization of polyanhydride nanoparticles. J Vis Exp. https://doi.org/10.3791/3967

Catalan J, Ilves M, Jarventaus H et al (2010) Genotoxic and immunotoxic effects of cellulose nanocrystals in vitro. Environ Mol Mutagen 405:391–405

Chavez-Santoscoy AV, Roychoudhury R, Pohl NLB et al (2012) Tailoring the immune response by targeting C-type lectin receptors on alveolar macrophages using "pathogen-like" amphiphilic polyanhydride nanoparticles. Biomaterials 33:4762–4772

Chen D, Kristensen D (2009) Opportunities and challenges of developing thermostable vaccines. Expert Rev Vaccines 8:547–557

Chen CH, Lin YL, Liu YK et al (2012) Liposome-based polymer complex as a novel adjuvant: enhancement of specifc antibody production and isotype switch. Int J Nanomed 7:607–621

Chen X, Liu Y, Wang L et al (2014) Enhanced humoral and cell-mediated immune responses generated by cationic polymer-coated PLA microspheres with adsorbed HBsAg. Mol Pharm 11:1772–1784

Chen N, Johnson MM, Collier MA et al (2018a) Tunable degradation of acetalated dextran microparticles enables controlled vaccine adjuvant and antigen delivery to modulate adaptive immune responses. J Control Release 273:147–159

Chen N, Gallovic MD, Tiet P et al (2018b) Investigation of tunable acetalated dextran microparticle platform to optimize M2e-based influenza vaccine efficacy. J Control Release 289:114–124

Clements CJ, Wesselingh SL (2005) Vaccine presentations and delivery technologies—what does the future hold? Expert Rev Vaccines 4:281–287

Clifton GT, Mittendorf EA, Peoples GE (2015) Adjuvant HER2/neu peptide cancer vaccines in breast cancer. Immunotherapy 7:1159–1168

Coffman RL, Sher A, Seder RA (2010) Vaccine adjuvants: putting innate immunity to work. Immunity 33:492–503

Coleman CM, Frieman MB (2014) Coronaviruses: important emerging human pathogens. J Virol 88:5209–5212

Coleman CM, Liu YV, Mu H et al (2014) Purified coronavirus spike protein nanoparticles induce coronavirus neutralizing antibodies in mice. Vaccine 32:3169–3174

Čolić M, Mihajlović D, Mathew A et al (2015) Cytocompatibility and immunomodulatory properties of wood based nanofibrillated cellulose. Cellulose 22:763–778

Collier MA, Junkins RD, Gallovic MD et al (2018) Acetalated dextran microparticles for codelivery of STING and TLR7/8 agonists. Mol Pharm 15:4933–4946

Cooper PD, Petrovsky N (2011) Delta inulin: a novel, immunologically active, stable packing structure comprising β-D-[2 → 1] poly(fructo-furanosyl) αd-glucose polymers. Glycobiology 21:595–606

Cooper PD, Barclay TG, Ginic-Markovic M, Petrovsky N (2013) The polysaccharide inulin is characterized by an extensive series of periodic isoforms with varying biological actions. Glycobiology 23:1164–1174

Cooper PD, Rajapaksha KH, Barclay TG et al (2015) Inulin crystal initiation via a glucose-fructose cross-link of adjacent polymer chains: atomic force microscopy and static molecular modelling. Carbohydr Polym 117:964–972

Danhier F, Ansorena E, Silva JM et al (2012) PLGA-based nanoparticles: an overview of biomedical applications. J Control Release 161:505–522

De Figueiredo P, Ficht TA, Rice-Ficht A et al (2015) Pathogenesis and immunobiology of brucellosis: review of Brucella-host interactions. Am J Pathol 185:1505–1517

De Geest BG, Willart MA, Lambrecht BN et al (2012) Surface-engineered polyelectrolyte multilayer capsules: synthetic vaccines mimicking microbial structure and function. Angew Chemie—Int Ed 51:3862–3866

De Titta A, Ballester M, Julier Z et al (2013) Nanoparticle conjugation of CpG enhances adjuvancy for cellular immunity and memory recall at low dose. Proc Natl Acad Sci U S A 110:19902–19907

Dehghan S, Kheiri MT, Abnous K et al (2018) Preparation, characterization and immunological evaluation of alginate nanoparticles loaded with whole inactivated influenza virus: dry powder formulation for nasal immunization in rabbits. Microb Pathog 115:74–85

Della Bella S, Bierti L, Presicce P et al (2007) Peripheral blood dendritic cells and monocytes are differently regulated in the elderly. Clin Immunol 122:220–228

Demento SL, Cui W, Criscione JM et al (2012) Role of sustained antigen release from nanoparticle vaccines in shaping the T cell memory phenotype. Biomaterials 33:4957–4964

Demicheli V, Jefferson T, Ferroni E, et al (2018) Vaccines for preventing influenza in healthy adults (Review)—summary of findings for the main comparison. Cochrane Database Syst Rev 1–254

Determan AS, Wilson JH, Kipper MJ et al (2006) Protein stability in the presence of polymer degradation products: consequences for controlled release formulations. Biomaterials 27:3312–3320

Dhakal S, Goodman J, Bondra K et al (2017a) Polyanhydride nanovaccine against swine influenza virus in pigs. Vaccine 35:1124–1131

Dhakal S, Hiremath J, Bondra K et al (2017b) Biodegradable nanoparticle delivery of inactivated swine influenza virus vaccine provides heterologous cell-mediated immune response in pigs. J Control Release 247:194–205

Dhakal S, Ghimire S, Renu S et al (2019) Evaluation of CpG-ODN-adjuvanted polyanhydride-based intranasal influenza nanovaccine in pigs. Vet Microbiol 237:108401

Di Cola E, Cantu L, Brocca P et al (2019) Novel O/W nanoemulsions for nasal administration: structural hints in the selection of performing vehicles with enhanced mucopenetration. Colloids Surf B Biointerfaces 183:110439

Didierlaurent AM, Laupèze B, Di Pasquale A et al (2017) Adjuvant system AS01: helping to overcome the challenges of modern vaccines. Expert Rev Vaccines 16:55–63
Draheim C, de Crécy F, Hansen S et al (2015) A design of experiment study of nanoprecipitation and nano spray drying as processes to prepare PLGA nano- and microparticles with defined sizes and size distributions. Pharm Res 32:2609–2624
Ekici S, Ilgin P, Butun S, Sahiner N (2011) Hyaluronic acid hydrogel particles with tunable charges as potential drug delivery devices. Carbohydr Polym 84:1306–1313
El-Sissi AF, Mohamed FH, Danial NM et al (2020) Chitosan and chitosan nanoparticles as adjuvant in local Rift Valley Fever inactivated vaccine. 3 Biotech 10:88
Facciolà A, Visalli G, Laganà P et al (2019) The new era of vaccines: the "nanovaccinology". Eur Rev Med Pharmacol Sci 23:7163–7182
Fan Y, Sahdev P, Ochyl LJ et al (2015) Cationic liposome-hyaluronic acid hybrid nanoparticles for intranasal vaccination with subunit antigens. J Control Release 208:121–129
Farhadian A, Dounighi NM, Avadi M (2015) Enteric trimethyl chitosan nanoparticles containing hepatitis B surface antigen for oral delivery. Hum Vaccines Immunother 11:2811–2818
Ferreira SA, Gama FM, Vilanova M (2013) Polymeric nanogels as vaccine delivery systems. Nanomed Nanotechnol Biol Med 9:159–173
Flach TL, Ng G, Hari A et al (2011) Alum interaction with dendritic cell membrane lipids is essential for its adjuvanticity. Nat Med 17:479–487
Franchi L, Núñez G (2008) The Nlrp3 inflammasome is critical for aluminum hydroxide-mediated IL-1β secretion but dispensable for adjuvant activity. Eur J Immunol 38:2085–2089
Galstyan A, Markman JL, Shatalova ES et al (2019) Blood–brain barrier permeable nano immunoconjugates induce local immune responses for glioma therapy. Nat Commun 10:1–13
Garlapati S, Facci M, Polewicz M et al (2009) Strategies to link innate and adaptive immunity when designing vaccine adjuvants. Vet Immunol Immunopathol 128:184–191
Ge F, Zhu L, Yang L et al (2018) The soluble and particulate form of alginates positively regulate immune response. Iran J Immunol 15:228–238
Gilbert BE, Patel N, Lu H et al (2018) Respiratory syncytial virus fusion nanoparticle vaccine immune responses target multiple neutralizing epitopes that contribute to protection against wild-type and palivizumab-resistant mutant virus challenge. Vaccine 36:8069–8078
Goodman JT, Ramirez JEVV, Boggiatto PM et al (2014) Nanoparticle chemistry and functionalization differentially regulates dendritic cell-nanoparticle interactions and triggers dendritic cell maturation. Part Part Syst Charact 31:1269–1280
Gratton SEA, Ropp PA, Pohlhaus PD et al (2008) The effect of particle design on cellular internalization pathways. Proc Natl Acad Sci U S A 105:11613–11618
Gu XG, Schmitt M, Hiasa A et al (1998) A novel hydrophobized polysaccharide/oncoprotein complex vaccine induces in vitro and in vivo cellular and humoral immune responses against HER2- expressing murine sarcomas. Cancer Res 58:3385–3390
Gu P, Wusiman A, Zhang Y et al (2019) Rational design of PLGA nanoparticle vaccine delivery systems to improve immune responses. Mol Pharm 16:5000–5012
Guo SH, Fu DW, Utupova A et al (2019) Applications of polymer-based nanoparticles in vaccine field. Nanotechnol Rev 8:143–155
Gupta RK, Rost BE, Relyveld E, Siber GR (1995) Adjuvant properties of aluminum and calcium compounds. Pharm Biotechnol 6:229–248
Hamdy S, Elamanchili P, Alshamsan A et al (2007) Responses by codelivery of ovalbumin and toll-like receptor ligand monophosphoryl lipid A in poly (D, L-lactic-co-glycolic acid) nanoparticles. J Biomed Mater Res A 81:652–662
Han J, Zhao D, Li D et al (2018) Polymer-based nanomaterials and applications for vaccines and drugs. Polymers (Basel) 10:1–14
Hanson CM, George AM, Sawadogo A, Schreiber B (2017) Is freezing in the vaccine cold chain an ongoing issue? A literature review. Vaccine 35:2127–2133
Haryono A, Salsabila K, Restu WK et al (2017) Effect of chitosan and liposome nanoparticles as adjuvant codelivery on the immunoglobulin G subclass distribution in a mouse model. J Immunol Res 2017:9125048

Haughney SL, Petersen LK, Schoofs AD et al (2013) Retention of structure, antigenicity, and biological function of pneumococcal surface protein A (PspA) released from polyanhydride nanoparticles. Acta Biomater 9:8262–8271

Haughney SL, Ross KA, Boggiatto PM et al (2014) Effect of nanovaccine chemistry on humoral immune response kinetics and maturation. Nanoscale 6:13770–13778

He Y, Park K (2016) Effects of the microparticle shape on cellular uptake. Mol Pharm 13:2164–2171

Highton AJ, Kojarunchitt T, Girardin A et al (2015) Chitosan hydrogel vaccine generates protective CD8 T cell memory against mouse melanoma. Immunol Cell Biol 93:634–640

Ho AWS, Prabhu N, Betts RJ et al (2011) Lung CD103 + dendritic cells efficiently transport influenza virus to the lymph node and load viral antigen onto MHC Class I for presentation to CD8 T cells. J Immunol 187:6011–6021

Hong F, Zhang F, Liu Y, Yan H (2017) DNA origami: Scaffolds for creating higher order structures. Chem Rev 117:12584–12640

Hosangadi D, Warmbrod KL, Martin EK et al (2020) Enabling emergency mass vaccination: innovations in manufacturing and administration during a pandemic. Vaccine 38:4167–4169

Hua S, de Matos MBC, Metselaar JM, Storm G (2018) Current trends and challenges in the clinical translation of nanoparticulate nanomedicines: pathways for translational development and commercialization. Front Pharmacol 9:790

Huntimer L, Wilson Welder JH, Ross K, et al (2013) Single immunization with a suboptimal antigen dose encapsulated into polyanhydride microparticles promotes high titer and avid antibody responses. J Biomed Mater Res—Part B Appl Biomater 101 B:91–98

Huntimer L, Ramer-Tait AE, Petersen LK et al (2013b) Evaluation of biocompatibility and administration site reactogenicity of polyanhydride-particle-based platform for vaccine delivery. Adv Healthc Mater 2:369–378

Huntimer LM, Ross KA, Darling RJ et al (2014) Polyanhydride nanovaccine platform enhances antigen-specific cytotoxic T cell responses. Technology 2:171–175

Igietseme JU, Eko FO, He Q, Black CM (2004) Antibody regulation of T-cell immunity: implications for vaccine strategies against intracellular pathogens. Expert Rev Vaccines 3:23–34

Ikeda M, Akagi T, Yasuoka T et al (2018) Characterization and analytical development for amphiphilic poly(γ-glutamic acid) as raw material of nanoparticle adjuvants. J Pharm Biomed Anal 150:460–468

Isaka M, Komiya T, Takahashi M et al (2004) Recombinant cholera toxin B subunit (rCTB) as a mucosal adjuvant enhances induction of diphtheria and tetanus antitoxin antibodies in mice by intranasal administration with diphtheria-pertussis-tetanus (DPT) combination vaccine. Vaccine 22:3061–3068

Ishii-Mizuno Y, Umeki Y, Onuki Y et al (2017) Improved sustained release of antigen from immunostimulatory DNA hydrogel by electrostatic interaction with chitosan. Int J Pharm 516:392–400

Jacob ST, Crozier I, Fischer WA et al (2020) Ebola virus disease. Nat Rev Dis Prim 6:13

Jahantigh D, Saadati M, Fasihi Ramandi M et al (2014) Novel intranasal vaccine delivery system by chitosan nanofibrous membrane containing N-terminal region of IpaD antigen as a nasal Shigellosis vaccine, studies in Guinea pigs. J Drug Deliv Sci Technol 24:33–39

Jain S, Tran TH, Amiji M (2015) Macrophage repolarization with targeted alginate nanoparticles containing IL-10 plasmid DNA for the treatment of experimental arthritis. Biomaterials 61:162–177

Jalah R, Patel V, Kulkarni V et al (2012) IL-12 DNA as molecular vaccine adjuvant increases the cytotoxic T cell responses and breadth of humoral immune responses in SIV DNA vaccinated macaques. Hum Vaccines Immunother 8:1620–1629

Jesus S, Soares E, Borchard G, Borges O (2018) Adjuvant activity of poly-ϵ-caprolactone/chitosan nanoparticles characterized by mast cell activation and IFN-γ and IL-17 production. Mol Pharm 15:72–82

Jin T, Tsao E (2013) Powder vaccines for pulmonary delivery. In: Flower D, Perrie Y (eds) Immunomic discovery of adjvuants and candidate subunit vaccines. Springer Science + Business Media, New York

Jindal AB (2017) The effect of particle shape on cellular interaction and drug delivery applications of micro- and nanoparticles. Int J Pharm 532:450–465

Jones CH, Chen M, Ravikrishnan A et al (2015) Mannosylated poly(beta-amino esters) for targeted antigen presenting cell immune modulation. Biomaterials 37:333–344

Joshi VB, Geary SM, Salem AK (2013) Biodegradable particles as vaccine delivery systems: size matters. AAPS J 15:85–94

Junkins RD, Gallovic MD, Johnson BM et al (2018) A robust microparticle platform for a STING-targeted adjuvant that enhances both humoral and cellular immunity during vaccination. J Control Release 270:1–13

Kageyama S, Kitano S, Hirayama M et al (2008) Humoral immune responses in patients vaccinated with 1-146 HER2 protein complexed with cholesteryl pullulan nanogel. Cancer Sci 99:601–607

Kang I, Hong MS, Nolasco H et al (2004) Age-associated change in the frequency of memory CD4+ T cells impairs long term CD4+ T cell responses to influenza vaccine. J Immunol 173:673–681

Kasturi SP, Skountzou I, Albrecht RA et al (2011) Programming the magnitude and persistence of antibody responses with innate immunity. Nature 470:543–547

Kelly SM, Mitra A, Mathur S, Narasimhan B (2020) Synthesis and characterization of rapidly degrading polyanhydrides as vaccine adjuvants. ACS Biomater Sci Eng 6:265–276

Kensil CR, Kammer R (2010) Expert opinion on investigational drugs: foreword. Expert Opin Investig Drugs 19:1475–1482

Kim MG, Park JY, Shon Y et al (2014) Nanotechnology and vaccine development. Asian J Pharm Sci 9:227–235

Kipper MJ, Shen E, Determan A, Narasimhan B (2002) Design of an injectable system based on bioerodible polyanhydride microspheres for sustained drug delivery. Biomaterials 23:4405–4412

Kitano S, Kageyama S, Nagata Y et al (2006) HER2-specific T-cell immune responses in patients vaccinated with truncated HER2 protein complexed with nanogels of cholesteryl pullulan. Clin Cancer Res 12:7397–7405

Klasse PJ (2014) Neutralization of virus infectivity by antibodies: old problems in new perspectives. Adv Biol 2014:1–24

Kokate RA, Thamake SI, Chaudhary P et al (2015) Enhancement of anti-tumor effect of particulate vaccine delivery system by "bacteriomimetic" CpG functionalization of poly-lactic-co-glycolic acid nanoparticles. Nanomedicine 10:915–929

Kokate RA, Chaudhary P, Sun X et al (2016) Rationalizing the use of functionalized poly-lactic-co-glycolic acid nanoparticles for dendritic cell-based targeted anticancer therapy. Nanomedicine 11:479–494

Kong WH, Sung DK, Kim H et al (2016) Self-adjuvanted hyaluronate—antigenic peptide conjugate for transdermal treatment of muscular dystrophy. Biomaterials 81:93–103

Kool M, Soullié T, Van Nimwegen M et al (2008) Alum adjuvant boosts adaptive immunity by inducing uric acid and activating inflammatory dendritic cells. J Exp Med 205:869–882

Kumar S, Tummala H (2013) Development of soluble inulin microparticles as a potent and safe vaccine adjuvant and delivery system. Mol Pharm 10:1845–1853

Kumar S, Kesharwani SS, Kuppast B et al (2016) Discovery of inulin acetate as a novel immune-active polymer and vaccine adjuvant: synthesis, material characterization, and biological evaluation as a toll-like receptor-4 agonist. J Mater Chem B 4:7950–7960

Kumar S, Kesharwani SS, Kuppast B et al (2017) Pathogen-mimicking vaccine delivery system designed with a bioactive polymer (inulin acetate) for robust humoral and cellular immune responses. J Control Release 261:263–274

Kumari A, Yadav SK, Yadav SC (2010) Biodegradable polymeric nanoparticles based drug delivery systems. Colloids Surf B Biointerfaces 75:1–18

Kumru OS, Joshi SB, Smith DE et al (2014) Vaccine instability in the cold chain: mechanisms, analysis and formulation strategies. Biologicals 42:237–259

Lampe AT, Farris EJ, Ballweg MD et al (2019) Non-complexed chitosan acts as an adjuvant in an influenza A virus protein vaccine. J Immunol 202(139):14

Lau Y-F, Wright AR, Subbarao K (2012) The contribution of systemic and pulmonary immune effectors to vaccine-induced protection from H5N1 influenza virus infection. J Virol 86:5089–5098

Lavigne MV, Castro M, Andino J, Manghi M (2004) Alternative diphtheria, tetanus and whooping cough immunization schedule to evoke a Th2 tetanus and a Th1 pertussis immune response. Microbes Infect 6:481–484

Lee BY, Bartsch SM, Mvundura M et al (2015) An economic model assessing the value of microneedle patch delivery of the seasonal influenza vaccine. Vaccine 33:4727–4736

Lee BY, Wedlock PT, Haidari LA et al (2017) Economic impact of thermostable vaccines. Vaccine 35:3135–3142

Leleux J, Roy K (2013) Micro and nanoparticle-based delivery systems for vaccine immunotherapy: an immunological and materials perspective. Adv Healthc Mater 2:72–94

Leroux-Roels G (2010) Unmet needs in modern vaccinology. Adjuvants to improve the immune response. Vaccine 28:C25–C36

Li X, Anton N, Arpagaus C et al (2010) Nanoparticles by spray drying using innovative new technology: the Büchi nano spray dryer B-90. J Control Release 147:304–310

Li P, Luo Z, Liu P et al (2013) Bioreducible alginate-poly(ethylenimine) nanogels as an antigen-delivery system robustly enhance vaccine-elicited humoral and cellular immune responses. J Control Release 168:271–279

Li P, Asokanathan C, Liu F et al (2016) PLGA nano/micro particles encapsulated with pertussis toxoid (PTd) enhances Th1/Th17 immune response in a murine model. Int J Pharm 513:183–190

Liu Q, Chen X, Jia J et al (2015) PH-Responsive poly(D, L-lactic-co-glycolic acid) nanoparticles with rapid antigen release behavior promote immune response. ACS Nano 9:4925–4938

Liu C, Chu X, Yan M et al (2018) Encapsulation of Poly I: C and the natural phosphodiester CpG ODN enhanced the efficacy of a hyaluronic acid-modified cationic lipid-PLGA hybrid nanoparticle vaccine in TC-1-grafted tumors. Int J Pharm 553:327–337

Lloyd J, Cheyne J (2017) The origins of the vaccine cold chain and a glimpse of the future. Vaccine 35:2115–2120

Locatelli L, Cadamuro M, Spirli C et al (2017) A STING-activating nanovaccine for cancer immunotherapy. Hepatology 63:965–982

Lopac SK, Torres MP, Wilson-Welder JH et al (2009) Effect of polymer chemistry and fabrication method on protein release and stability from polyanhydride microspheres. J Biomed Mater Res—Part B Appl Biomater 91:938–947

Lu J, Liu X, Liao YP et al (2017) Nano-enabled pancreas cancer immunotherapy using immunogenic cell death and reversing immunosuppression. Nat Commun 8:1811

Luo M, Wang H, Wang Z et al (2017) A STING-activating nanovaccine for cancer immunotherapy. Nat Nanotechnol 12:648–654

MacLennan ICM, Toellner K-M, Cunningham AF et al (2003) Extrafollicular antibody responses. Immunol Rev 194:8–18

Magnusson SE, Reimer JM, Karlsson KH et al (2013) Immune enhancing properties of the novel Matrix-M™ adjuvant leads to potentiated immune responses to an influenza vaccine in mice. Vaccine 31:1725–1733

Malik A, Gupta M, Gupta V et al (2018) Novel application of trimethyl chitosan as an adjuvant in vaccine delivery. Int J Nanomedicine 13:7959–7970

Manish M, Rahi A, Kaur M et al (2013) A single-dose PLGA encapsulated protective antigen domain 4 nanoformulation protects mice against *Bacillus anthracis* spore challenge. PLoS ONE 8:e61885

Mann AJ, Noulin N, Catchpole A et al (2014) Intranasal H5N1 vaccines, adjuvanted with chitosan derivatives, protect ferrets against highly pathogenic influenza intranasal and intratracheal challenge. PLoS ONE 9:e93761

Martínez Rivas CJ, Tarhini M, Badri W et al (2017) Nanoprecipitation process: from encapsulation to drug delivery. Int J Pharm 532:66–81

Masomian M, Ahmad Z, Gew LT (2020) Development of next generation *Streptococcus pneumoniae* vaccines conferring broad protection. Vaccines 8:132

McAdams D, Chen D, Kristensen D (2012) Spray drying and vaccine stabilization. Expert Rev Vaccines 11:1211–1219

McComb RC, Martchenko M (2016) Neutralizing antibody and functional mapping of *Bacillus anthracis* protective antigen—the first step toward a rationally designed anthrax vaccine. Vaccine 34:13–19

McGill JL, Kelly S, Narasimhan B, et al (2017) Efficacy and immunogenicity of a novel polyanhydride nanovaccine for use against RSV infection in the neonate. J Immunol 198:225.4 LP-225.4

McGill JLL, Kelly SMM, Kumar P et al (2018) Efficacy of mucosal polyanhydride nanovaccine against respiratory syncytial virus infection in the neonatal calf. Sci Rep 8:1–15

McNeela EA, Mills KHG (2001) Manipulating the immune system: humoral versus cell-mediated immunity. Adv Drug Deliv Rev 51:43–54

Mehta NR, Wurz GT, Burich RA et al (2012) L-BLP25 vaccine plus letrozole induces a T H1 immune response and has additive antitumor activity in MUC1-expressing mammary tumors in mice. Clin Cancer Res 18:2861–2871

Mittendorf EA, Clifton GT, Holmes JP et al (2014) Final report of the phase I/II clinical trial of the E75 (nelipepimut-S) vaccine with booster inoculations to prevent disease recurrence in high-risk breast cancer patients. Ann Oncol 25:1735–1742

Mittendorf EA, Ardavanis A, Symanowski J et al (2016a) Primary analysis of a prospective, randomized, single-blinded phase II trial evaluating the HER2 peptide AE37 vaccine in breast cancer patients to prevent recurrence. Ann Oncol 27:1241–1248

Mittendorf EA, Ardavanis A, Litton JK et al (2016b) Primary analysis of a prospective, randomized, single-blinded phase II trial evaluating the HER2 peptide GP2 vaccine in breast cancer patients to prevent recurrence. Oncotarget 7:66192–66201

Mkrtichyan M, Ghochikyan A, Movsesyan N et al (2011) Immunostimulant adjuvant patch enhances humoral and cellular immune responses to DNA immunization. Bone 23:1–7

Mohri K, Nishikawa M, Takahashi Y, Takakura Y (2014) DNA nanotechnology-based development of delivery systems for bioactive compounds. Eur J Pharm Sci 58:26–33

Mohri K, Kusuki E, Ohtsuki S et al (2015) Self-assembling DNA dendrimer for effective delivery of immunostimulatory CpG DNA to immune cells. Biomacromol 16:1095–1101

Montomoli E, Piccirella S, Khadang B et al (2011) Current adjuvants and new perspectives in vaccine formulation. Expert Rev Vaccines 10:1053–1061

Moon S, Shin EC, Noh YW, Lim YT (2015) Evaluation of hyaluronic acid-based combination adjuvant containing monophosphoryl lipid A and aluminum salt for hepatitis B vaccine. Vaccine 33:4762–4769

Moran HBT, Turley JL, Andersson M, Lavelle EC (2018) Immunomodulatory properties of chitosan polymers. Biomaterials 184:1–9

Moreno-Mendieta S, Guillén D, Hernández-Pando R et al (2017) Potential of glucans as vaccine adjuvants: a review of the α-glucans case. Carbohydr Polym 165:103–114

Mosafer J, Sabbaghi AH, Badiee A et al (2019) Preparation, characterization and in vivo evaluation of alginate-coated chitosan and trimethylchitosan nanoparticles loaded with PR8 influenza virus for nasal immunization. Asian J Pharm Sci 14:216–221

Moyano DF, Goldsmith M, Solfiell DJ et al (2012) Nanoparticle hydrophobicity dictates immune response. J Am Chem Soc 134:3965–3967

Muruganandah V, Sathkumara HD, Navarro S, Kupz A (2018) A systematic review: the role of resident memory T cells in infectious diseases and their relevance for vaccine development. Front Immunol 9:1574

Nam J, Son S, Park KS et al (2019) Cancer nanomedicine for combination cancer immunotherapy. Nat Rev Mater 4:398–414

Narasimhan B, Goodman JT, Vela Ramirez JE (2016) Rational design of targeted next-generation carriers for drug and vaccine delivery. Annu Rev Biomed Eng 18:25–49

Nashchekina YA, Raydan M (2018) Noninvasive penetration of 5 nm hyaluronic acid molecules across the epidermal barrier (in vitro) and its interaction with human skin cells. Skin Res Technol 24:129–134

Neamtu I, Rusu AG, Diaconu A et al (2017) Basic concepts and recent advances in nanogels as carriers for medical applications. Drug Deliv 24:539–557

Nevagi RJ, Skwarczynski M, Toth I (2019a) Polymers for subunit vaccine delivery. Eur Polym J 114:397–410

Nevagi RJ, Dai W, Khalil ZG et al (2019b) Self-assembly of trimethyl chitosan and poly(anionic amino acid)-peptide antigen conjugate to produce a potent self-adjuvanting nanovaccine delivery system. Bioorganic Med Chem 27:3082–3088

Nicolete R, Santos DFD, Faccioli LH (2011) The uptake of PLGA micro or nanoparticles by macrophages provokes distinct in vitro inflammatory response. Int Immunopharmacol 11:1557–1563

Niikura K, Matsunaga T, Suzuki T et al (2013) Gold nanoparticles as a vaccine platform: influence of size and shape on immunological responses in vitro and in vivo. ACS Nano 7:3926–3938

Nunberg JH, Doyle MV, York SM, York CJ (1989) Interleukin 2 acts as an adjuvant to increase the potency of inactivated rabies virus vaccine. Proc Natl Acad Sci U S A 86:4240–4243

Ohmit SE, Petrie JG, Malosh RE et al (2013) Influenza vaccine effectiveness in the community and the household. Clin Infect Dis 56:1363–1369

Olafsdottir TA, Lingnau K, Nagy E, Jonsdottir I (2009) IC31®, a two-component novel adjuvant mixed with a conjugate vaccine enhances protective immunity against pneumococcal disease in neonatal mice. Scand J Immunol 69:194–202

Oleszycka E, McCluskey S, Sharp FA et al (2018) The vaccine adjuvant alum promotes IL-10 production that suppresses Th1 responses. Eur J Immunol 48:705–715

Ostrom QT, Bauchet L, Davis FG et al (2014) The epidemiology of glioma in adults: a state of the science review. Neuro Oncol 16:896–913

Paliwal R, Babu RJ, Palakurthi S (2014) Nanomedicine scale-up technologies: feasibilities and challenges. Ageing Int 15:1527–1534

Pati R, Shevtsov M, Sonawane A (2018) Nanoparticle vaccines against infectious diseases. Front Immunol 9:2224

Pawar D, Goyal AK, Mangal S et al (2010) Evaluation of mucoadhesive PLGA microparticles for nasal immunization. AAPS J 12:130–137

Pearse MJ, Drane D (2005) ISCOMATRIX® adjuvant for antigen delivery. Adv Drug Deliv Rev 57:465–474

Pennock ND, White JT, Cross EW et al (2013) T cell responses: naive to memory and everything in between. AJP Adv Physiol Educ 37:273–283

Pereira MM, Raposo NRB, Brayner R et al (2013) Cytotoxicity and expression of genes involved in the cellular stress response and apoptosis in mammalian fibroblast exposed to cotton cellulose nanofibers. Nanotechnology 24:075103

Peres C, Matos AI, Conniot J et al (2017) Poly(lactic acid)-based particulate systems are promising tools for immune modulation. Acta Biomater 48:41–57

Petersen LK, Ramer-Tait AE, Broderick SR et al (2011) Activation of innate immune responses in a pathogen-mimicking manner by amphiphilic polyanhydride nanoparticle adjuvants. Biomaterials 32:6815–6822

Petersen LK, Phanse Y, Ramer-Tait AE et al (2012) Amphiphilic polyanhydride nanoparticles stabilize bacillus anthracis protective antigen. Mol Pharm 9:874–882

Petrovsky N, Cooper PD (2011) Carbohydrate-based immune adjuvants. Expert Rev Vaccines 10:523–537

Phanse Y, Carrillo-Conde BR, Ramer-Tait AE et al (2013) Functionalization of polyanhydride microparticles with di-mannose influences uptake by and intracellular fate within dendritic cells. Acta Biomater 9:8902–8909

Phanse Y, Lueth P, Ramer-Tait AE et al (2016) Cellular internalization mechanisms of polyanhydride particles: implications for rational design of drug delivery vehicles. J Biomed Nanotechnol 12:1544–1552

Phanse Y, Carrillo-Conde BR, Ramer-Tait AE et al (2017) Functionalization promotes pathogen-mimicking characteristics of polyanhydride nanoparticle adjuvants. J Biomed Mater Res—Part A 105:2762–2771

Piao Z, Akeda Y, Takeuchi D et al (2014) Protective properties of a fusion pneumococcal surface protein A (PspA) vaccine against pneumococcal challenge by five different PspA clades in mice. Vaccine 32:5607–5613

Plotkin S, Robinson JM, Cunningham G et al (2017) The complexity and cost of vaccine manufacturing—an overview. Vaccine 35:4064–4071

Quandt D, Rothe K, Baerwald C, Rossol M (2015) GPRC6A mediates alum-induced Nlrp3 inflammasome activation but limits Th2 type antibody responses. Sci Rep 5:1–12

Rajput MKS, Kesharwani SS, Kumar S et al (2018) Dendritic cell-targeted nanovaccine delivery system prepared with an immune-active polymer. ACS Appl Mater Interfaces 10:27589–27602

Ramirez JEV, Roychoudhury R, Habte HH et al (2014) Carbohydrate-functionalized nanovaccines preserve HIV-1 antigen stability and activate antigen presenting cells. J Biomater Sci Ed 25:1387–1406

Ramirez JEV, Boggiatto PM, Wannemuehler MJ, Narasimhan B (2017) Polyanhydride nanoparticle interactions with host serum proteins and their effects on bone marrow derived macrophage activation. ACS Biomater Sci Eng 3:160–168

Rao JP, Geckeler KE (2011) Polymer nanoparticles: preparation techniques and size-control parameters. Prog Polym Sci 36:887–913

Rayahin JE, Buhrman JS, Zhang Y et al (2015) High and low molecular weight hyaluronic acid differentially influence macrophage activation. ACS Biomater Sci Eng 1:481–493

Reed SG, Orr MT, Fox CB (2013) Key roles of adjuvants in modern vaccines. Nat Med 19:1597–1608

Renu S, Feliciano-Ruiz N, Ghimire S et al (2020) Poly(I:C) augments inactivated influenza virus-chitosan nanovaccine induced cell mediated immune response in pigs vaccinated intranasally. Vet Microbiol 242:108611

Riedel S (2005) Plague: From natural disease to bioterrorism. Baylor Univ Med Cent Proc 18:116–124

Rose F, Wern JE, Gavins F et al (2018) A strong adjuvant based on glycol-chitosan-coated lipid-polymer hybrid nanoparticles potentiates mucosal immune responses against the recombinant *Chlamydia trachomatis* fusion antigen CTH522. J Control Release 271:88–97

Ross KA, Loyd H, Wu W et al (2014) Structural and antigenic stability of H5N1 hemagglutinin trimer upon release from polyanhydride nanoparticles. J Biomed Mater Res—Part A 102:4161–4168

Ross KA, Lyod H, Wu W et al (2015) Hemagglutinin-based polyanhydride nanovaccines against H5N1 infuenza elicit protective virus neutralizing titers and cell-mediated immunity. Int J Nanomedicine 10:229–243

Ross K, Senapati S, Alley J et al (2019) Single dose combination nanovaccine provides protection against influenza A virus in young and aged mice. Biomater Sci 7:809–821

Rothan HA, Byrareddy SN (2020) The epidemiology and pathogenesis of coronavirus disease (COVID-19) outbreak. J Autoimmun 109:102433

Roy J, Adili R, Kulmacz R et al (2016) Development of poly unsaturated fatty acid derivatives of aspirin for inhibition of platelet functions. J Pharmacol Exp Ther 359:134–141

Saboo S, Tumban E, Peabody J et al (2016) Optimized formulation of a thermostable spray-dried virus-like particle vaccine against human papillomavirus. Mol Pharm 13:1646–1655

Saravia J, Chapman NM, Chi H (2019) Helper T cell differentiation. Cell Mol Immunol 16:634–643

Sarei F, Dounighi NM, Zolfagharian H et al (2013) Alginate nanoparticles as a promising adjuvant and vaccine delivery system. Indian J Pharm Sci 75:442–449

Sarti F, Perera G, Hintzen F et al (2011) In vivo evidence of oral vaccination with PLGA nanoparticles containing the immunostimulant monophosphoryl lipid A. Biomaterials 32:4052–4057

Saxena M, Sabado RL, La Mar M et al (2019) Poly-ICLC, a TLR3 agonist, induces transient innate immune responses in patients with treated HIV-infection: a randomized double-blinded placebo controlled trial. Front Immunol 10:1–12

Schmid K, Arpagaus C, Friess W (2011) Evaluation of the nano spray dryer B-90 for pharmaceutical applications. Pharm Dev Technol 16:287–294

Semete B, Booysen LIJ, Kalombo L et al (2010) In vivo uptake and acute immune response to orally administered chitosan and PEG coated PLGA nanoparticles. Toxicol Appl Pharmacol 249:158–165

Sharma G, Valenta DT, Altman Y et al (2010) Polymer particle shape independently influences binding and internalization by macrophages. J Control Release 147:408–412

Sharma S, Parmar A, Kori S, Sandhir R (2015) PLGA-based nanoparticles: a new paradigm in biomedical applications. Trends Anal Chem 80:30–40

Shen E, Pizsczek R, Dziadul B, Narasimhan B (2001) Microphase separation in bioerodible copolymers for drug delivery. Biomaterials 22:201–210

Shen L, Higuchi T, Tubbe I et al (2013) A trifunctional dextran-based nanovaccine targets and activates murine dendritic cells, and induces potent cellular and humoral immune responses in vivo. PLoS ONE 8:e80904

Shen L, Li J, Liu Q et al (2018) Local blockade of interleukin 10 and C-X-C Motif chemokinelLigand 12 with nano-delivery promotes antitumor response in murine cancers. ACS Nano 12:9830–9841

Shi J, Kantoff PW, Wooster R, Farokhzad OC (2017) Cancer nanomedicine: progress, challenges and opportunities. Nat Rev Cancer 17:20–37

Shi N-Q, Zhou J, Walker J et al (2020) Microencapsulation of luteinizing hormone-releasing hormone agonist in poly (lactic-co-glycolic acid) microspheres by spray-drying. Front Pharmacol 321:756–772

Shim BS, Park SM, Quan JS et al (2010) Intranasal immunization with plasmid DNA encoding spike protein of SARS-coronavirus/polyethylenimine nanoparticles elicits antigen-specific humoral and cellular immune responses. BMC Immunol 11:65

Shima F, Akagi T, Uto T, Akashi M (2013) Manipulating the antigen-specific immune response by the hydrophobicity of amphiphilic poly($\gamma$-glutamic acid) nanoparticles. Biomaterials 34:9709–9716

Silva JM, Videira M, Gaspar R et al (2013) Immune system targeting by biodegradable nanoparticles for cancer vaccines. J Control Release 168:179–199

Silva AL, Soema PC, Slutter B et al (2016) PLGA particulate delivery systems for subunit vaccines: linking particle properties to immunogenicity. Hum Vaccin Immunother 12:1056–1069

Singh D, Somani VK, Aggarwal S, Bhatnagar R (2015) PLGA (85: 15) nanoparticle based delivery of rL7/L12 ribosomal protein in mice protects against Brucella abortus 544 infection: a promising alternate to traditional adjuvants. Mol Immunol 68:272–279

Singh RS, Kaur N, Rana V, Kennedy JF (2017) Pullulan: a novel molecule for biomedical applications. Carbohydr Polym 171:102–121

Song Y, Zhou Y, Van Drunen Littel-Van Den Hurk S, Chen L (2014) Cellulose-based polyelectrolyte complex nanoparticles for DNA vaccine delivery. Biomater Sci 2:1440–1449

Sosnik A, Seremeta KP (2015) Advantages and challenges of the spray-drying technology for the production of pure drug particles and drug-loaded polymeric carriers. Adv Colloid Interface Sci 223:40–54

Stephens L, Ross KA, Hartwig SM et al (2019) Robust immune response and protection generated by a polyanhydride-based nanoparticle vaccine utilizing the RSV prefusion F protein and/or M protein. J Immunol 202:139.11 LP-139.11

Stephens L, Ross KA, Hartwig SM et al (2020) Polyanhydride-based RSV prefusion F nanoparticle vaccine establishes lasting protection mediated by humoral and cellular immune responses. J Immunol 204:166.3 LP-166.3

Stieneker F, Kersten G, Van BL et al (1995) Comparison of 24 different adjuvants for inactivated HIV-2 split whole virus as antigen in mice. Induction of titres of binding antibodies and toxicity of the formulations. Vaccine 13:45–53

Stupp R, Mason WP, van den Bent MJ et al (2005) Radiotherapy plus concomitant and adjuvant Temozolomide for glioblastoma. N Engl J Med 352:987–996

Sui Z, Chen Q, Fang F et al (2010) Cross-protection against influenza virus infection by intranasal administration of M1-based vaccine with chitosan as an adjuvant. Vaccine 28:7690–7698

Swanson KA, Rainho-Tomko JN, Williams ZP et al (2020) A respiratory syncytial virus (RSV) F protein nanoparticle vaccine focuses antibody responses to a conserved neutralization domain. Sci Immunol 5:1–12

Tamargo RJ, Myseros JS, Epstein JI et al (1993) Interstitial chemotherapy of the 9L gliosarcoma: controlled release polymers for drug delivery in the brain. Cancer Res 53:329–333

Tamayo I, Irache JM, Mansilla C et al (2010) Poly(Anhydride) nanoparticles act as active Th1 adjuvants through toll-like receptor exploitation. Clin Vaccine Immunol 17:1356–1362

Tang A, Chen Z, Cox KS et al (2019) A potent broadly neutralizing human RSV antibody targets conserved site IV of the fusion glycoprotein. Nat Commun 10:1–13

Tao W, Gill HS (2015) M2e-immobilized gold nanoparticles as influenza A vaccine: role of soluble M2e and longevity of protection. Vaccine 33:2307–2315

Thomas C, Rawat A, Hope-Weeks L, Ahsan F (2011) Aerosolized PLA and PLGA nanoparticles enhance humoral, mucosal and cytokine responses to hepatitis B vaccine. Mol Pharm 8:405–415

Tomić S, Kokol V, Mihajlović D et al (2016) Native cellulose nanofibrills induce immune tolerance in vitro by acting on dendritic cells. Sci Rep 6:1–14

Torres MP, Wilson-Welder JH, Lopac SK et al (2011) Polyanhydride microparticles enhance dendritic cell antigen presentation and activation. Acta Biomater 7:2857–2864

Torres FG, Troncoso OP, Pisani A et al (2019) natural polysaccharide nanomaterials: an overview of their immunological properties. Int J Mol Sci 20

Trent A, Ulery BD, Black MJ et al (2015) Peptide amphiphile micelles self-adjuvant group A Streptococcal vaccination. AAPS J 17:380–388

Truong-Dinh Tran T, Ha-Lien Tran P, Tu Nguyen K, Tran V-T (2016) Nano-precipitation: preparation and application in the field of pharmacy. Curr Pharm Des 22:2997–3006

Tzeng SY, Guarecuco R, McHugh KJ et al (2016) Thermostabilization of inactivated polio vaccine in PLGA-based microspheres for pulsatile release. J Control Release 233:101–113

Ulery BD, Phanse Y, Sinha A et al (2009) Polymer chemistry influences monocytic uptake of polyanhydride nanospheres. Pharm Res 26:683–690

Ulery BD, Petersen LK, Phanse Y et al (2011) Rational design of pathogen-mimicking amphiphilic materials as nanoadjuvants. Sci Rep 1:1–9

Urbanavicius D, Alvarez T, Such GK et al (2018) The potential of nanoparticle vaccines as a treatment for cancer. Mol Immunol 98:2–7

Uto T, Toyama M, Nishi Y et al (2013) Uptake of biodegradable poly(γ-glutamic acid) nanoparticles and antigen presentation by dendritic cells in vivo. Results Immunol 3:1–9

Uto T, Akagi T, Akashi M, Baba M (2015) Induction of potent adaptive immunity by the novel polyion complex nanoparticles. Clin Vaccine Immunol 22:578–585

Vartiainen J, Pöhler T, Sirola K et al (2011) Health and environmental safety aspects of friction grinding and spray drying of microfibrillated cellulose. Cellulose 18:775–786

Vela Ramirez JE (2015) Design of polyanhydride-based targeted nanovaccines against HIV. ProQuest Diss Theses 259

Vela Ramirez JE, Tygrett LT, Hao J et al (2016) Polyanhydride nanovaccines induce germinal center B cell formation and sustained serum antibody responses. J Biomed Nanotechnol 12:1303–1311

Vesely MD, Kershaw MH, Schreiber RD, Smyth MJ (2011) Natural innate and adaptive immunity to cancer. Annu Rev Immunol 29:235–271

Wadajkar AS, Dancy JG, Hersh DS et al (2017) Tumor-targeted nanotherapeutics: overcoming treatment barriers for glioblastoma. Wiley Interdiscip Rev Nanomed Nanobiotechnol 9:1–24

Wafa EI, Geary SM, Goodman JT et al (2017) The effect of polyanhydride chemistry in particle-based cancer vaccines on the magnitude of the anti-tumor immune response. Acta Biomater 50:417–427

Wafa EI, Geary SM, Ross KA et al (2019) Single dose of a polyanhydride particle-based vaccine generates potent antigen-specific antitumor immune responses. J Pharmacol Exp Ther 370:855–863

Wagner DA, Kelly SM, Petersen AC et al (2019) Single-dose combination nanovaccine induces both rapid and long-lived protection against pneumonic plague. Acta Biomater 100:326–337

Wagner-Muñiz DA, Haughney SL, Kelly SM et al (2018) Room temperature stable PspA-based nanovaccine induces protective immunity. Front Immunol 9:325

Wait SD, Prabhu RS, Burri SH et al (2015) Polymeric drug delivery for the treatment of glioblastoma. Neuro Oncol 17:ii9–ii23

Wen ZS, Xu YL, Zou XT, Xu ZR (2011) Chitosan nanoparticles act as an adjuvant to promote both Th1 and Th2 immune responses induced by ovalbumin in mice. Mar Drugs 9:1038–1055

Wen Y, Mu L, Shi Y (2016) Immunoregulatory functions of immune complexes in vaccine and therapy. EMBO Mol Med 8:1120–1133

WHO (2008) Anthrax in humans and animals. World Organ Anim Heal 219

Wicki A, Witzigmann D, Balasubramanian V, Huwyler J (2015) Nanomedicine in cancer therapy: challenges, opportunities, and clinical applications. J Control Release 200:138–157

Wilson-Welder JHH, Torres MPP, Kipper MJJ et al (2009) Vaccine adjuvants: current challenges and future approaches. J Pharm Sci 98:1278–1316

Wisniewski R (2015) Spray drying technology review. Int Conf Environ Syst 1–9

Wu YL, Park K, Soo RA et al (2011) INSPIRE: a phase III study of the BLP25 liposome vaccine (L-BLP25) in Asian patients with unresectable stage III non-small cell lung cancer. BMC Cancer 11:430

Wu TYH, Singh M, Miller AT et al (2014) Rational design of small molecules as vaccine adjuvants. Sci Transl Med 6:160

Wu NC, Zost SJ, Thompson AJ et al (2017) A structural explanation for the low effectiveness of the seasonal influenza H3N2 vaccine. PLoS Pathog 13:1–17

Wusiman A, Gu P, Liu Z et al (2019) Cationic polymer modified PLGA nanoparticles encapsulating alhagi honey polysaccharides as a vaccine delivery system for ovalbumin to improve immune responses. Int J Nanomedicine 14:3221–3234

Xia Y, Fan Q, Hao D et al (2015) Chitosan-based mucosal adjuvants: sunrise on the ocean. Vaccine 33:5997–6010

Xu J, Dai W, Wang Z et al (2011) Intranasal vaccination with chitosan-DNA nanoparticles expressing pneumococcal surface antigen A protects mice against nasopharyngeal colonization by *Streptococcus pneumoniae*. Clin Vaccine Immunol 18:75–81

Xu JH, Dai WJ, Chen B, Fan XY (2015) Mucosal immunization with PsaA protein, using chitosan as a delivery system, increases protection against acute otitis media and invasive infection by *Streptococcus pneumoniae*. Scand J Immunol 81:177–185

Yang HW, Ye L, Guo XD et al (2017) Ebola vaccination using a DNA vaccine coated on PLGA-PLL/γPGA nanoparticles administered using a microneedle patch. Adv Healthc Mater 6

Yoshizaki Y, Yuba E, Sakaguchi N et al (2017) pH-sensitive polymer-modified liposome-based immunity-inducing system: effects of inclusion of cationic lipid and CpG-DNA. Biomaterials 141:272–283

Yue ZG, Wei W, Lv PP et al (2011) Surface charge affects cellular uptake and intracellular trafficking of chitosan-based nanoparticles. Biomacromol 12:2440–2446

Zacharias ZR, Ross KA, Hornick EE et al (2018) Polyanhydride nanovaccine induces robust pulmonary B and T cell immunity and confers protection against homologous and heterologous influenza A virus infections. Front Immunol 9:1953

Zehrung D, Jarrahian C, Giersing B, Kristensen D (2017) Exploring new packaging and delivery options for the immunization supply chain. Vaccine 35:2265–2271

Zhang YQ, Tsai YC, Monie A et al (2010) Carrageenan as an adjuvant to enhance peptide-based vaccine potency. Vaccine 28:5212–5219

Zhang Z, Tongchusak S, Mizukami Y et al (2011) Biomaterials induction of anti-tumor cytotoxic T cell responses through PLGA-nanoparticle mediated antigen delivery. Biomaterials 32:3666–3678

Zhang L, Zhu G, Mei L et al (2015) Self-assembled DNA immunonanoflowers as multivalent CpG nanoagents. ACS Appl Mater Interfaces 7:24069–24074

Zhang W, An M, Xi J, Liu H (2017a) Targeting CpG adjuvant to lymph node via dextran conjugate enhances antitumor immunotherapy. Bioconjug Chem 28:1993–2000

Zhang Q, Hu M, Xu L et al (2017b) Effect of edible fungal polysaccharides on improving influenza vaccine protection in mice. Food Agric Immunol 28:981–992

Zhao Y, Wang Y, Ran F et al (2017) A comparison between sphere and rod nanoparticles regarding their in vivo biological behavior and pharmacokinetics. Sci Rep 7:1–11

Zhu C, Shoji Y, McCray S et al (2014) Stabilization of HAC1 influenza vaccine by spray drying: Formulation development and process scale-up. Pharm Res 31:3006–3018

Zost SJ, Parkhouse K, Gumina ME et al (2017) Contemporary H3N2 influenza viruses have a glycosylation site that alters binding of antibodies elicited by egg-adapted vaccine strains. Proc Natl Acad Sci U S A 114:12578–12583

# Virus-Like Particle Vaccines Against Respiratory Viruses and Protozoan Parasites

**Ki-Back Chu and Fu-Shi Quan**

## Contents



K.-B. Chu
Department of Biomedical Science, Graduate School, Kyung Hee University, Seoul 02447, Republic of Korea

F.-S. Quan (✉)
26, Kyungheedae-ro, Dongdaemun-gu, Seoul 02447, Republic of Korea
e-mail: fsquan@khu.ac.kr

Department of Medical Zoology, School of Medicine, Kyung Hee University, Seoul 02447, Republic of Korea

Medical Research Center for Bioreaction To Reactive Oxygen Species and Biomedical Science Institute, School of Medicine, Kyung Hee University, Graduate school, Seoul 02447, Republic of Korea

Current Topics in Microbiology and Immunology (2021) 433: 77–106
https://doi.org/10.1007/82_2021_232

Published Online: 3 March 2021

**Abstract** The field of vaccinology underwent massive advances over the past decades with the introduction of virus-like particles (VLPs), a supra-molecular nanoparticle vaccine platform that resembles viral structures without the ability to replicate in hosts. This innovative approach has been remarkably effective, as evidenced by its profound immunogenicity and safety. These highly desirable intrinsic properties enabled their further development as vaccines against a multitude of diseases. To date, several VLP-based vaccines have already been commercialized and many more are undergoing clinical evaluation prior to FDA approval. However, efficacious vaccines against a plethora of pathogens are still lacking, which imposes a tremendous socioeconomic burden and continues to threaten public health throughout the globe. This is especially the case for several respiratory pathogens and protozoan parasites. In this review, we briefly describe the fundamentals of VLP vaccines and the unique properties that enable these to be such valuable vaccine candidates and summarize current advances in VLP-based vaccines targeting respiratory and parasitic diseases of global importance.

## 1 Introduction

Massive advances to the biomedical science have been made in the twenty-first century, resulting in the emergence of novel molecular technologies such as the construction of virus-like particles (VLPs). Vaccinology in particular harnessed the full power of these innovative approaches and application of these concepts invigorated vaccine development and research against infectious pathogens. Importantly, the virus-like particle (VLP) vaccines are of keen interest for vaccine design strategy due to several intrinsic properties that render VLPs more effective than the traditional vaccines. Traditionally, vaccine design strategies were predominantly based on using inactivated or live-attenuated pathogens but several drawbacks to these production methods have fueled and accelerated the development of alternative strategies. Attenuated vaccines are derived from weakened pathogens, and though these vaccines are capable of inducing long-term immunity, the possibility of the pathogen reverting to its virulent wild-type form remains a challenge to be addressed. Applying heat or formaldehyde to inactivate pathogens for use as vaccine components is much safer than the attenuated vaccines, but undesirable side effects including exuberant inflammation have necessitated the need for a safer vaccine (Gause et al. 2017). Moreover, vaccines developed using these methods were only effective against pathogens with relatively simple pathogenesis, whereas such strategies were deemed less effective against pathogens involving complex life cycles or host–pathogen interactions (Jennings and Bachmann 2008). Subunit vaccines expressing carbohydrate and protein antigens are much safer than these traditional vaccines, but their safety aspect comes at the cost of diminished immunogenicity (Moyle and Toth 2013). The VLPs are an exemplification of efficacious vaccines possessing favorable parameters of both subunit and traditional vaccines, notably the high immunogenicity of traditional vaccines and the safety aspect of subunit vaccines as represented by their replication-deficient nature.

The first VLP vaccines were constructed in the 1980s against hepatitis B virus (HBV). Only years later, the first VLP-based vaccine was approved for clinical use and many others followed. Over the past decades, with enormous improvements to genetic engineering technologies, VLP vaccine design strategies have made giant strides to achieve several developmental milestones (Plotkin 2014; Lua et al. 2014). Currently, a diverse array of VLP formulation methods have been developed and computational tools are actively being employed to enhance vaccine immunogenicity. VLP-based vaccines are being extensively studied as prophylactic tools against many microbial organisms. These include VLP-based vaccines against hepatitis A virus (HAV), hepatitis B virus (HBV), and human papillomavirus (HPV), all of which are commercially available while many other are undergoing clinical evaluation (Kushnir et al. 2012). Despite these endeavors, vaccines against etiological agents responsible for some of the most important diseases are still lacking. Viruses and parasites are especially of concern for causing a high rate of mortality throughout the globe. Influenza and respiratory syncytial virus (RSV) infections have been viral-borne respiratory diseases of importance and VLP-based vaccines are currently lacking for these pathogens or vaccine in general in the latter case. A vaccine against neglected but important tropical diseases such as malaria and toxoplasmosis is also direly needed, but decades of developmental effort have remained unfruitful. In this chapter, we will describe the VLP technology, and review the current progress in several major infectious diseases lacking an efficacious VLP vaccine, and highlight global attempts to develop one from both pre-clinical and clinical standpoints.

## 2 Virus-Like Particles (VLPs) and Properties Enabling Immunity Induction

VLPs are self-assembled highly immunogenic particles that mimic the morphology of the natural virus particle while lacking the genetic components required for replication. These particles are organized into repetitive elements, usually comprising a foreign antigen and a base platform. This repetitive aspect is of particular importance as mounting an efficient antibody response to foreign antigen epitope requires the foreign epitope to be expressed repetitively on a particle surface (Fehr et al. 1998; Bachmann et al. 1993). Importantly, the highly repetitive and densely presented antigens serve as a pathogen-associated structural pattern (PASP) that facilitates direct cross-linking with the B cell receptor without T cell assistance (Mohsen et al. 2017a; Brune and Howarth 2018).

Another factor that contributes to their application in vaccines is their small size. Ranging anywhere between 20 and 200 nm, these particles can be easily trafficked and they rapidly accumulate in the lymph nodes (Manolova et al. 2008). Interestingly, upon immunizing mice with bacteria-derived VLPs, the VLPs were found to have accumulated in the popliteal lymph nodes of the immunized mice in less than 10 min (Mohsen et al. 2017b). In the case of porcine parvovirus VLPs, it took less than 2 h for half of the entire murine splenic dendritic cells to be presented with the antigen from VLPs (Morón et al. 2002). In addition to their rapid drainage into the lymph nodes for enhancing B cell response, VLP vaccines also induce cytotoxic T cell response via major histocompatibility complex cross-presentation (Brune and Howarth 2018; Gomes et al. 2017). Moreover, particulate vaccines such as VLPs generally possess charged surfaces with intrinsic properties enabling receptor interaction, which allows enhanced interaction with APCs such as DCs than solubilized proteins (Bachmann and Jennings 2010).

## 2.1 VLP Production System

The VLP production and expression systems and the pitfalls associated with each system have been extensively described elsewhere (Kushnir et al. 2012; Zeltins 2013; Charlton Hume and Lua 2017; Donaldson et al. 2015). In brief, VLPs can be produced in a variety of prokaryotic and eukaryotic cells which includes bacterial, yeast, mammalian, insect, or plant cells. Each of these production strategies has its own limitations and all of these production methods require the presence of viruses for proper icosahedral structured VLP production. These viruses include the hepatitis B virus (HBV), bacteriophages MS2, Qβ, P22, the plant viruses cowpea chlorotic mottle virus (CCMV) and cowpea mosaic virus (CPMV), and the baculovirus system in insects.

*Escherichia coli*, along with other members in the family *Enterobacteriacea*, is highly sought after for VLP production. Infection of these bacteria using the bacteriophages MS2, Qβ, and the *Salmonella typhimurium* P22 is required to drive the VLP production process (Rohovie et al. 2017). Although bacteria-derived recombinant proteins are widely utilized for their low production cost and high yields, their use for VLP production is not recommended which can be attributed to the absence of mammalian-like post-translational modifications (PTM), and formation of inclusion bodies. Bacteria-derived VLPs are also unable to form proper disulfide bonds, have issues with protein stability and solubility, and the presence of endotoxins renders their use unacceptable for clinical trials (Zeltins 2013). The yeast VLP expression system is a bit more complex than bacteria, which shares many of the features of bacteria-derived VLPs, with the presence of notable differences in PTM especially glycosylation (Kushnir et al. 2012). However, the overall yield is much lower than bacteria-derived VLPs.

Numerous transgenic plants have been reported to date as potential source of VLP production (Bragard et al. 2000; Huang et al. 2006; Mason et al. 1996). Plant cells can produce VLPs at a higher yield at a low cost, but as with above, antigen glycosylation and other PTM processes are issues that need to be overcome (Lomonossoff and D'Aoust 2016). There are also stability issues such as antigens being degraded during in vivo delivery, which limits its applicability (Roldão et al. 2010). Plant-based system using *Nicotiana* sp. has been of concern due to the toxic alkaloid build-up. However, this issue has been overcome by crossing two transgenic *Nicotiana* spp. (Ling et al. 2012). VLPs produced using mammalian cells may be the optimal choice, as folding and PTM of complex proteins may not be an issue. Nevertheless, the mammalian cell approach is the most expensive method and requires stringent quality control, as they are prone to passenger virus infections (Plotkin et al. 2017). Other factors hindering its use are slow growth and low yield (Donaldson et al. 2015).

The baculovirus is a DNA virus that naturally infects insect cells. Although they can be transduced using mammalian cells, baculoviruses are incapable of replicating in mammalian cells and cytotoxicity is virtually none. Because these are non-pathogenic to humans, these have been widely utilized in the field of biomedical science (Luo et al. 2013). Baculovirus expression system using insect cells is quite frequently used and is a powerful tools for producing VLP vaccines (Fig. 1). Insect cells enable high production yields in relatively short-time frames and are much safer and easier to handle than mammalian cells, since several known human pathogens are absent. Commonly used cell lines are the insect cells belonging to the Lepidoptera family, notably Sf9 or Sf21 cells from *Spodoptera frugiperda* and High Five™ cells from *Trichoplusia ni* (Liu et al. 2013; Puente-Massaguer et al. 2020). Though this production method still remains as a favorable method of choice, there are limitation to this production method. First and foremost, co-production of enveloped baculovirus particles and other contaminants has been reported to impair vaccine efficacy. Consequently, these have to be removed through chemical inactivation or via downstream processing procedures that further affects the immunogenicity of VLPs (Rueda et al. 2000; Hervas-Stubbs et al. 2007). As with plant cells, insect cell-derived VLPs using baculovirus also have PTM-associated issues most notably the ones involving *N*-glycans. Although transgenic insect cell lines are currently being researched to overcome this problem, this still remains a work in progress and further developments are required (Brune and Howarth 2018).

One of the latest methods for modular VLP vaccine design is based on the Tag/Catcher system introduced by Zakeri et al. (2012). The fundamental concept underlying this technology is the formation of an intramolecular isopeptide bond, which is frequently found in the collagen-binding adhesin (Cna) domains of the cell surface proteins belonging to numerous Gram-positive bacteria (Kang and Baker 2011). These isopeptide bonds were also observed in the CnaB2 domain of fibronectin-binding protein (FbaB) in *Streptococcus pyogenes*, where the lysine and the aspartate residues spontaneously formed a covalent bond of phenomenal strength that were resilient to diverse environmental conditions (Zakeri et al. 2012). Based on this finding, genetic fusion of the antigen of interest to the SpyTag and attaching this component to the SpyCatcher-expressing core protein would enable easy construction of VLPs.

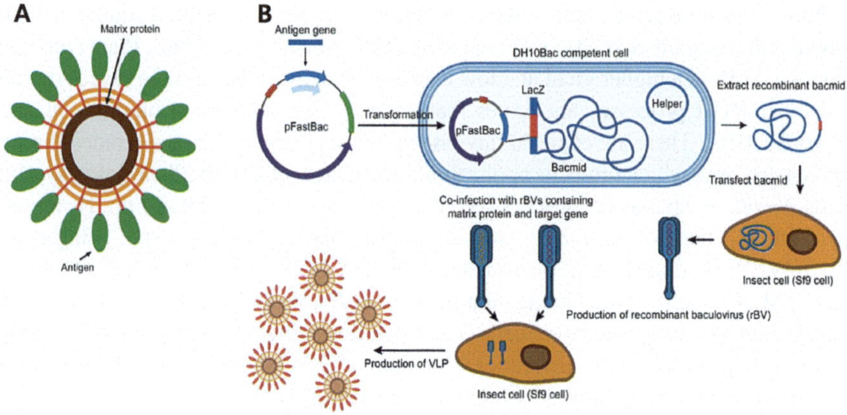


**Fig. 1** Structure and production schematic of insect cell-based VLP vaccines. **a** A structural representation of a typical virus-like particle vaccine, which illustrates the matrix protein surrounded by the phospholipid bilayer and the repeated expression of antigens on the surface. **b** A VLP vaccine production schematic using the baculovirus expression system. Briefly, the antigen gene of interest needs to be cloned into the pFastBac™ vector and subsequently transformed into the DH10Bac competent cell. The helper plasmid and the baculovirus shuttle vector allows transposition to occur within the DH10Bac cells, resulting in a recombinant bacmid DNA. This bacmid DNA is then transfected into the *Spodoptera frugiperda* (Sf9) insect cell for recombinant baculovirus production. Finally, co-transfecting the recombinant baculovirus expressing the matrix protein with the recombinant baculovirus containing the antigen of interest into the insect cell results in the production of VLPs that can be used as vaccines post-purification

## 2.2 *Advantages and Disadvantages of VLPs Compared to Other Vaccine Platforms*

There are multiple aspects that render VLP-based vaccines more effective than traditional vaccines. Since VLPs are completely devoid of pathogenic genomic material, replication is impossible and are therefore safe for use (Chackerian 2007). Furthermore, as discussed earlier, their structural resemblance to viruses and the geometric arrangement of highly repetitive antigens enables potent immune response induction (Fehr et al. 1998; Bachmann et al. 1993). Compared to other vaccines, protective efficacies are much greater in VLP vaccines than traditional vaccines. VLPs can be rapidly produced using various expression systems with robust, scalable production systems that are highly cost-efficient (Zhang et al. 2014). Currently, commercially available VLP-based vaccines are generally produced using the baculoviruses expression system or yeast expression system. One advantage of using the insect baculovirus system is that they are unable to replicate in mammalian cells, which further supports the notion of the VLPs being safe for use (Hervas-Stubbs et al. 2007). With regards to immunogenicity, VLPs generally do not require adjuvant usage as these are highly immunogenic by themselves. However, by supplementing these vaccines with adjuvants, the vaccine efficacy can be further enhanced.

There are several down-sides to the VLP-based vaccines. First and foremost, their cost of production is rather expensive compared to other vaccine platforms. Secondly, other factors such as downstream purification processes are required, as the presence of host cellular impurities can be detrimental during clinical trials (Effio and Hubbuch 2015). Besides cost, the purification steps also add a time penalty to vaccine production. There is also potential of side effects associated with VLP use. Although rare, HBV VLPs produced using yeast were reported to have elicited anaphylaxis in several patients (DiMiceli et al. 2006).

## 3 Influenza Virus

The influenza virus belongs to the *Orthomyxoviridae* family, which consists of the genera *Alphainfluenzavirus, Betainfluenzavirus, Gammainfluenzavirus,* and *Deltainfluenzavirus* along with few others. Its annual impact places a tremendous economic burden, with direct medical costs reported to be $10.4 billion in the USA alone (Molinari et al. 2007). Avian influenza is also of importance as its outbreaks can lead to the culling of millions of poultry that devastates the poultry industry and animal health. Circulating avian influenza virus has resulted in the culling of 400 million birds and the economic loss incurred was estimated to be around $20 billion USD (Short et al. 2015). With the continuous threat from annual influenza, an effective influenza vaccine that can be rapidly manufactured has become a priority. In 2013, the seasonal influenza vaccine FluBlok was approved for clinical use. This recombinant vaccine, unlike the previous vaccines manufactured from embryonated eggs, was produced using the insect cell Sf9 with high immunogenicity (Liu et al. 2013). With several advantages of the insect cell-based vaccine system, other studies are following suit and investigators are starting to shy away from the traditional egg-based vaccines which have been the standard for over 50 years (Cox and Hollister 2009). To date, multitudes of influenza VLP vaccine studies have been conducted and the overall findings were generally consistent, with vaccines inducing potent antibody responses and protection.

### *3.1 Influenza Vaccine Components*

There are several genes of antigenicity that are frequently utilized as vaccine components. Among the genes encoded by the influenza virus, the most well-characterized proteins are the hemagglutinin (HA) and neuraminidase (NA). Apart from these, the virus also encodes several other structural and non-structural proteins such as the nucleoprotein (NP), matrix protein (M1), ion channel protein (M2), non-structural proteins (NS1, NS2), and several polymerase complexes (PB1, PB2, PA) (Jang and Seong 2020). Nevertheless, the highly mutating nature of influenza surface glycoproteins has become nothing more than a nuisance for vaccine development. Currently, the main targets for the commercialized influenza virus vaccines are the globular head domains of the HA. This has become problematic and research focusing on other areas that are more conserved has been emerging. Several examples of these conserved regions are the stalk domain of HA, the ectodomain of the ion channel (M2e), domains of NA, and even internal structural proteins such as M1 and NP (Krammer et al. 2018).

## *3.2 Seasonal Influenza VLP Vaccines*

Although strategies for VLP construction differ between research groups, the overall result seems to be more or less similar. Nitrocellulose membrane-based filtration was used to concentrate the insect-derived VLPs expressing the HA and NA of the highly pathogenic avian influenza (HPAI) A/ Korea/ Mallard/W452/ 2014) H5N8. This filtration process did not affect the immunogenicity of the vaccines, as immunized mice mounted strong virus-specific antibody responses (Park and Song 2017). Plant cell-based VLPs are also being investigated as a potential influenza vaccine candidate. The plant *Nicotiana benthamiana* has been widely used to generate stable and highly immunogenic VLPs. Their immunization in mice even elicited protection in old mice that were experiencing multiple co-morbidities, indicating its potential for conferring protection in the elderly (Lindsay et al. 2018; Hodgins et al. 2019). A cocktail VLP vaccine constituted of H1, H3, H5, or H7 HAs conferred broad protection against several subtypes of influenza A virus, which include the 1918 H1, 1957 H2, as well as the avian H5, H6, H7, H10, and H11 subtypes (Schwartzman et al. 2015). NA VLP generated using the A/PR/8/34 (H1N1) successfully protected mice against both homologous and heterosubtypic lethal challenge infection (Quan et al. 2012).

Another strategy is utilizing computationally optimized broadly reactive antigen (COBRA). Using this method, the codon-optimized HA VLPs were produced in mammalian cells which were used to immunize mice and ferrets. Vaccinated animals were completely protected from lethal challenge infection with the clade 2.2 H5N1 virus A/Mongolia/Whooper Swan/244/2005 (Giles and Ross 2011). The key difference here is that ferrets were involved rather than simply testing the VLP vaccine efficacy in mice. This is crucial since the murine immune system is not an accurate depiction of human immunity. Through these results, the potential for VLPs as a seasonal influenza vaccine is quite encouraging.

## 3.3 *Pandemic Influenza VLP Vaccine*

Influenza A virus pandemics have occurred several times within the last century. Since its first report in 1918 with the H1N1 influenza virus, three other pandemics occurred in the years 1957, 1968, and 2009 involving the H2N2, H3N2, and H1N1pdm, each respectively (Viboud et al. 2020). These pandemics, which occurred in the twentieth and twenty-first century incurred millions of deaths (Ryu and Cowling 2020). Currently, developing pandemic influenza vaccines is of concern since the efficacy of seasonal influenza vaccines against novel pandemic influenza A virus is extremely low or negligible at best. This development process is further hampered through the time required for development upon correct matching of circulating pandemic vaccine strain, which can take up to six months (Krammer and Palese 2015; Krietsch Boerner 2020).

Several VLP vaccines against pandemic 2009 H1N1 have been constructed using the baculovirus expression vector system, and the conferred immunity enabled protection in both mice and ferret models (Quan et al. 2010; Song et al. 2011). HA VLPs containing the transmembrane domain of H3N2 exhibited protection against H7N9 challenge, a strain that has been perceived to have numerous outbreaks in countries such as China (Qin et al. 2018). VLPs targeting avian influenza have also been generated and they were proven to be efficacious in both murine and ferret models (Song et al. 2010). Replacing the influenza M1 protein with another protein from a different virus could be a potential vaccine design strategy. Using the gag protein isolated from bovine immunodeficiency virus (Bgag), VLPs expressing HAs of H5N1, H7N9, H9N2, and H10N8 have been constructed (Tretyakova et al. 2016). With multiple HA subtypes being expressed, this vaccine could potentially serve as a first line of defense upon avian influenza pandemic after efficacy confirmation in animal models.

## 3.4 *Universal Influenza VLP Vaccine*

Designing a universal influenza vaccine that broadly protects against various subtypes is highly desired, but the process has remained challenging simply due to the nature of the virus. RNA viruses such as the influenza virus use its own RNA polymerase for replication, which lacks fidelity and is quite error-prone. Consequently, this frequent build-up of mutations results in an antigenic drift that renders the host immune system ineffective since antibodies are unable to recognize the viral antigenic site (Zambon 1999). This is particularly important as exemplified during the 2014/2015 influenza season. At the time, the most prevalent strain circulating around the globe for H3N2 was A/Texas/50/2012, but by the end of the year, antigenic drift caused a transition from this strain to A/Switzerland/9,715,293/2013 being the most prevalent strain. Consequently, this resulted in suboptimal vaccine efficacy estimated to be 18% for the 2014/2015 season (Berlanda Scorza et al. 2016).

Numerous attempts have been made to produce a universal influenza vaccine conferring long-lasting cross-protection against various subtypes using the more conserved antigens aforementioned. Bivalent heterologous vaccine regimen involving immunization with both DNA vaccine and VLPs conferred broad protection against a variety of influenza subtypes. Vaccination using this method ensured the production of antibodies that bound to HAs from both group 1 and group 2 (Jiang et al. 2017).

Apart from this combinatorial vaccination approach, the VLP vaccines containing the heterologous tandem repeats of the M2 extracellular domain (M2e) were found to be highly cross-protective (Kim and Song 2013; Kim et al. 2014, 2017). Supplementing several vaccines, whether it be live-attenuated or pandemic split vaccines, with the M2e heterologous tandem repeat VLPs enhanced the efficacy of vaccines (Lee et al. 2019a). The efficacies of these M2e VLPs were further assessed in ferrets and chickens, which better reflect human and avian influenza pathogenesis, and the results appeared promising (Music et al. 2016; Song et al. 2016). The development of influenza VLP vaccines and its protective efficacies have been extensively reviewed elsewhere (Quan et al. 2020).

## 3.5 VLP Vaccines for Clinical Trials

To date, numerous clinical trials have been conducted using VLPs and most of the clinical results seem to point toward the conclusion that VLPs are highly immunogenic and safe for use in humans. An alum-adjuvanted plant-derived VLP vaccine expressing the HA protein of H5N1 (A/Indonesia/5/05) successfully induced cross-protection in ferrets and humans. Notably, in the clinical phase I trial, the VLPs were found to be safe and highly immunogenic as indicated by hemagglutination inhibiton (HI) and microneutralization responses, thereby paving the way for further evaluation. With regard to side effects, most of the symptoms such as pain at the immunization site were mild to self-limiting at best (Landry et al. 2010). In an FDA-approved clinical trial phase I/II study, immunization with the clade 2.1 H5N1 VLP expressing HA, NA, and M1 induced the production of antibodies, which conferred cross-protection against the other clade 2 subtypes (Khurana et al. 2011). VLP vaccines against other avian influenza strains, notably H7N9, have been on-going. Adjuvanted-VLP vaccines expressing H7 antigen promoted enhanced production of neutralizing antibodies compared to unadjuvanted VLPs, even at a lower dose (Chung et al. 2015).

With the start of the 2009 H1N1 influenza outbreak occurring in Mexico, there was a dire need for a vaccine but a vaccine became available in the USA, five months after the start of the outbreak. Phase II clinical trial results for the VLP vaccine against the 2009 H1N1 (A/California/04/2009) revealed that high HI titers were induced in the vast majority of the participants and were overall well-tolerated (López-Macías et al. 2011). Similar findings were observed from clinical trials conducted in Singapore using *E. coli*-derived VLP vaccine against H1N1 (A/

California/07/2009). This bacteria-derived VLP vaccine successfully elicited influenza-specific $CD4^+$ and $CD8^+$ T cell responses following vaccination, along with a plethora of cytokines (Low et al. 2014; Skibinski et al. 2018). Though the results of this study appear promising, clinical evaluation involving human challenge infection with the influenza virus is required to confirm its protective efficacy. Recently, the immunogenicity and safety aspect of a quadrivalent VLP vaccine has been confirmed in the phase II clinical trial. The quadrivalent VLPs expressing the HA proteins of H1N1 (A/California/07/2009), H3N2 (A/Victoria/361/11), the Victoria lineage B/Brisbane/60/08, and the Yamagata lineage B/Massachusetts/02/2012 were produced in plants. Immunizing the participants with 30 μg of this VLPs ensured strong induction of both humoral and cellular immune responses and will be subjected to phase III clinical trial in the near future (Pillet et al. 2019).

Immunizing participants with the insect-derived 2009 H1N1 VLPs ensured durable antibody response that persisted up to 25 months post-immunization, and re-vaccinating these participants with the trivalent inactivated influenza vaccine enhanced the protective antibody levels (Valero-Pacheco et al. 2016). Several clinical trials assessing the efficacies of both H1 (A/California/07/2009) and H5 VLPs (A/Indonesia/5/05) have reported similar results. Immunization with H5 VLPs mounted durable antibody and cellular immune responses, while subjects immunized with H1-expressing VLP produced in plants induced stronger $CD4^+$ T cell responses than patients receiving placebo or the trivalent inactivated vaccine (Landry et al. 2014). A clinical evaluation assessing the humoral immune response to glycosylated H1 and H5 VLPs revealed that their immunization did not exacerbate pre-existing allergic reactions nor incurred allergies or other hypersensitivity symptoms (Ward et al. 2014).

## 4 Respiratory Syncytial Virus (RSV)

RSV was first isolated in 1956 but designing a vaccine for this virus has remained a monumental task. In 1966, clinical trial using formalin-inactivated RSV (FI-RSV) vaccine sensitized the patients and eventually led to the death of two children (Efstathiou et al. 2020). Since this medical failure nearly half a century ago, there was a hesitancy for conducting RSV vaccine-related studies. However, with the resurgence of novel vaccine design platforms enabling improvements to previously utilized vaccination strategies, the effort to design an RSV vaccine has been re-ignited. To date, RSV infection still remains as a major risk factor for lower respiratory tract infection (LRTI)-related hospitalizations for infants. Statistically, RSV accounts for approximately 28% of all acute LRTI cases and 22% of all LRTI-related mortality in children (Shi et al. 2017). In the year 2017, the global cost estimate for RSV-associated LRTI management in children under five years of age was determined to be approximately $5.2 billion USD (Eichinger et al. 2020; Zhang et al. 2020). For these reasons, developing a successful RSV vaccine has become a global necessity.

## 4.1 RSV VLP Vaccine Antigen Components

RSV is a negative-sense RNA virus consisting of ten genes that encode a total of 11 different proteins. The transmembrane proteins that are expressed on the viral surface are the fusion glycoprotein (F), attachment glycoprotein (G), and the small hydrophobic protein (SH). Other internal proteins encoded by the virus include the nucleoprotein (N), RNA polymerase (L), phosphoprotein (P), matrix protein (M), the transcription factors (M2-1, M2-2), and non-structural proteins (NS1, NS2) Efstathiou et al. 2020; Collins et al. 2013). Of the three proteins expressed on the surface, F and G proteins are more extensively studied than the SH protein, as both of these are capable of inducing effective antibody response (Ha et al. 2020; Cane 2001). The SH protein has also been tested as an antigen candidate, but their immunogenicity upon natural infection with the RSV was deemed weak as indicated by the low immunoglobulin levels in both humans and murine models (Akerlind-Stopner et al. 1993; Schepens et al. 2014).

Although the F protein is more conserved in comparison to the G protein, both are capable of inducing effective antibody responses. Structural analysis results have revealed that the RSV F protein can exist in either pre-fusion or a post-fusion conformation. Of the two conformations, majority of the research was focused on the pre-fusion (Pre-F) form since most of the neutralizing activity was directed here rather than the post-fusion form. One antigenic site of interest in the Pre-F protein is the "site Ø," which was shown to elicit the strongest neutralizing antibody response against RSV (Ngwuta et al. 2015; Frey et al. 2020). Evidently, the antibody response induced through site Ø was as much as 100 times more potent than those induced upon treatment with the commercialized monoclonal antibody palivizumab (McLellan et al. 2013). Despite these insightful findings, it was revealed that site Ø was the least conserved site on the F protein in the A and B subtypes, while the regions outside of site Ø retained 96% similarity across subtypes (McLellan et al. 2013).

## 4.2 Experimental VLP-Based Vaccines for RSV

Preliminary RSV VLP vaccines tested in animal models appear to be promising. Thus far, various VLPs have been produced presenting F, G, or the two combined using numerous platforms. Chimeric VLPs were among the first VLP vaccines developed against RSV. Chimeric VLPs constructed using the newcastle disease virus (NDV) and RSV F or G proteins have demonstrated interesting results. The ectodomain of RSV G protein fused to the NDV HN protein stimulated immune response comparable to those induced by UV-inactivated RSV vaccine without exerting much side effects. Similar findings were also observed from chimeric VLP constructed using the RSV F protein fused to the NDV F protein, with the absence of pulmonary pathologies (Murawski et al. 2010; McGinnes et al. 2011). Recent investigation using this chimeric VLP elucidated that mutations stabilizing the structure of RSV F protein in its Pre-F conformation induce greater titer of antibody responses to cope with subsequent RSV infection (McGinnes Cullen et al. 2015). Immunizing mice that were previously exposed to RSV with this chimeric VLP vaccine resulted in a drastic increase in protective antibody titers (McGinnes Cullen et al. 2019). Most recently, these chimeric VLPs were used to assess the anti-RSV immunity and protective efficacy via maternal immunization in cotton rats. Immunized dams gave birth to pups that were also protected from RSV challenge, indicated by lessened pulmonary inflammation (Blanco et al. 2018).

Immunizing the mice twice with the either RSV F or G protein expressed on influenza matrix protein 1 (M1) conferred significant protection against RSV A2 challenge infection, with enhanced viral clearance being observed from RSV G VLP (Quan et al. 2011). In both mice and cotton rats, combining both RSV F and G protein elicited enhanced protection without undesirable side effects, whereas RSV G protein immunization alone exacerbated the pulmonary inflammatory damage incurred by RSV challenge infection (Lee et al. 2014; Hwang et al. 2017a). Several VLPs displaying tandem repeats of the RSV G protein suppressed pulmonary eosinophilia and viral titers in mice (Kim et al. 2018). VLPs constructed using this method have been applied with other vaccines and this heterologous immunization strategy appeared to be promising. Co-immunizing mice using these VLPs along with an RSV F DNA vaccine conferred durable antibody responses and protection while mitigating the inflammatory cell influx responsible for lung pathology (Hwang et al. 2014; Ko et al. 2015). In line with this notion, identical findings were also observed from cotton rats immunized using this strategy (Hwang et al. 2016). Interestingly, priming mice with RSV F VLPs then boosting with FI-RSV vaccine prevented the development of vaccine-enhanced respiratory diseases (Hwang et al. 2017b). Both active and passive immunization with the insect cell-derived recombinant RSV F VLPs conferred broad protection against RSV A and B subtypes in cotton rats (Raghunandan et al. 2014).

A comparative study assessing the efficacy of RSV F VLP to those of FI-RSV and live-attenuated RSV vaccines revealed interesting findings. In contrast to the latter two, RSV F VLP vaccine immunization elicited RSV neutralizing antibody response, $CD8\alpha^+$ and $CD103^+$ dendritic cells while minimizing the pulmonary damage incurred by RSV challenge infection (Kim et al. 2015). Similarly, while a similar degree of neutralizing antibody responses was observed from both RSV F VLPs and F soluble protein, immunization with the former prevented vaccine-enhanced respiratory diseases while the latter was associated with lung histopathology upon RSV challenge (Lee et al. 2017a). The efficacies of a VLP vaccine generated using the matrix protein of human metapneumovirus (hMPV) matrix protein (M) expressing Pre-F, Post-F, or both were assessed in BALB/c mice. Contrary to the effects observed from FI-RSV vaccine immunization, VLP vaccines suppressed pulmonary inflammation. Interestingly, VLP vaccine displaying both Pre-F and Post-F immunogens induced the highest level of neutralizing antibody response, as well as the Th1 response required for viral clearance (Cimica et al. 2016).

Immunizing cotton rats with mammalian cell-derived VLPs expressing the RSV F and G proteins elicited potent neutralizing antibody responses that enhanced viral clearance in both lower and upper respiratory tracts (Walpita et al. 2015). RSV VLPs expressing the various forms of F and G proteins have been produced using different platforms: through one VLP expressing M + P proteins and the other expressing M + M2-1 proteins (Ha et al. 2020). The F and G proteins, regardless of how they were presented as truncated form, full length, or peptide form, induced RSV F and G protein-specific antibody responses and lessened the lung viral titer upon challenge infection in mice (Ha et al. 2020). In cotton rats, pre-fusion F VLPs conferred significant protection against RSV in the offspring of immunized dams (Cullen et al. 2020). As with the influenza virus VLP vaccines, the protective efficacies of these aforementioned RSV VLP vaccines have been thoroughly reviewed elsewhere (Quan et al. 2020).

### *4.3 RSV Vaccines for Clinical Trials*

Clinical VLP-based RSV vaccine studies are extremely limited. The most advanced vaccine is the VLP vaccine manufactured from Novavax using the Sf9/rBV technology (Mazur et al. 2018). This VLP vaccine, which expresses the F protein of RSV, has recently completed the phase III clinical evaluation which involved the immunization of 4,636 healthy pregnant women. The pregnant women were intramuscularly immunized with a single dose of RSV F protein VLP vaccine and the infants were followed up to a year for LRTI and safety assessment. Overall, although the vaccines were safe and incurred minimal adverse events, the vaccine failed to meet its pre-specified success criterion (Madhi et al. 2020). Despite these results, the vaccines are promising and further improvements will be required prior to FDA approval. Apart from this vaccine, the most recent RSV VLP vaccine that will be undergoing phase I clinical trial is the synthetic VLP vaccine V-306 [ClinicalTrials.gov ID: NCT04519073].

## 5 Toxoplasmosis

The causative organism of toxoplasmosis is the protozoan parasite *Toxoplasma gondii* which belongs to the phylum *Apicomplexa*. It has been reported that virtually all animals, including but not limited to mammals and avians, are capable of being infected by this parasite (Black and Boothroyd 2000). To date, due to their ubiquitous nature, more than 1 billion individuals throughout the globe have been estimated to be infected with *T. gondii*. Furthermore, *T. gondii* infection is one of the most common parasite-associated food-borne illness which requires hospitalization (Pappas et al. 2009; Montazeri et al. 2020). Though majority of the infected individuals remain asymptomatic with mild flu-like symptoms, their presence can have devastating consequences which can be fatal at times. In particular, acquired immune deficiency syndrome patients and pregnant women are especially at risk of encephalitis and congenital toxoplasmosis. Their infection in livestock can result in abortion or stillbirth, thus inflicting substantial damage to the agricultural industry (Wang et al. 2019). To lessen the socioeconomic burden incurred by *T. gondii*, vaccines are urgently needed. Thus far, despite decades of research, clinical toxoplasmosis vaccine remains unavailable while a commercial live-attenuated vaccine is available for ovines (Dubey 2009).

Currently, a variety of strategies have been implemented to develop *T. gondii* vaccines for clinical use. Traditional vaccines based on inactivated parasites have failed to demonstrate efficacy in multitudes of the models tested to date, while live-attenuated vaccines still face safety concerns that renders them unacceptable for use in humans as attenuated strains can revert back to virulent wild type. An extensively large number of protein families belonging to *T. gondii* have been tested as potential vaccine candidates (Rezaei et al. 2019). The major antigens used in these recombinant subunit vaccines include members of the dense granule antigens (GRA), microneme proteins (MIC), rhoptry proteins (ROP), and the surface antigens (SAG) (Li and Zhou 2018). Vaccines expressing these antigens were capable of eliciting sufficient antibody responses but failed to confer full protection against pathogens, and as such, further development is required. DNA vaccines have also proven to be effective, but the immune response induced in animals has been elicited to varying degree. Particularly, immunization with DNA vaccine in large animals such as ovines have demonstrated subpar efficacy (Li and Zhou 2018). In comparison to the aforementioned strategies, VLP-based *T. gondii* vaccines have proven to be highly efficacious, often conferring full protection against lethal challenge infection. Further efficacy testing in higher-order eukaryotes is still needed, but the future of *T. gondii* VLP vaccines appears promising.

## *5.1* Toxoplasma Gondii *Antigens and VLP Vaccines*

Numerous *T. gondii* vaccine studies have been conducted using DNA, recombinant protein subunit, or live-attenuated vaccines. Contrary to this, VLP-based *T. gondii* vaccine studies are extremely limited in number and all of them are experimental vaccines that have not been clinically assessed. Although different vaccine platforms were used, these earlier vaccine studies have delineated the antigenicity of numerous parasitic proteins that may serve as potential vaccine candidate antigens. Among the numerous antigen candidates reported to date, studies were predominantly focused on proteins involved in the parasite's motility, replication, host cell attachment, or invasion. Examples of these include the *T. gondii* GRA, MIC, ROP, SAG, and many others such as the apical membrane antigen (AMA) (Wang et al. 2019).

One of the earliest VLP-based *T. gondii* vaccines was developed using the inner membrane complex expressed on the spherical influenza virus M1 protein (Lee et al. 2016a, b). Using an identical format, several other vaccines expressing the rhoptry proteins were constructed and their efficacies were assessed in mice (Lee et al. 2017b). Other antigens that are involved in the host cell invasion by the *T. gondii* were expressed using VLPs and their efficacies were assessed (Kim et al. 2020). The efficacies of these VLPs were further enhanced by expressing multiple antigens (Lee et al. 2018a, b; Kang et al. 2020,2019). Chimeric VLP vaccines only conferred partial protection in mice against the virulent *T. gondii* RH strain (Guo et al. 2019). Overall, all of the aforementioned VLP vaccines were tested in mice with varying degrees of efficacies (Table 1). Therefore, further evaluation and improvements to the *T. gondii* VLP vaccines are required prior to clinical trials.

## 6 Malaria

Malaria still remains as one of the deadliest diseases affecting humans. In the year 2019, the number of malaria incidences was 229 million cases worldwide, and the number of deaths was estimated to be 409,000, with 67% of the deaths occurring in children under five years of age (World Health Organization 2020). Decades of effort have enabled a considerable reduction in global malaria disease burden and the mortality associated with it between the years 2000 and 2017 (Weiss et al. 2019). However, recent meta-analysis data has revealed that the overall rates of congenital and neonatal malaria in the endemic regions were approximately 40% and 12%, respectively (Danwang et al. 2020). The high prevalence associated with the disease requires additional efforts and investment to overcome. From a socioeconomic standpoint, malaria is of utmost importance as the incidence of malaria was inversely correlated with economic growth. Notably, industrial growth was much slower in malaria-endemic areas despite having the same labor intensity (Sarma et al. 2019). Even in areas where malaria is not endemic, the economic impact of malaria cannot be neglected. From the years 2000 to 2014, malaria-associated total hospital costs and total charges were roughly $176 million and $555 million in the USA, respectively (Khuu et al. 2019). With these issues underscoring the importance of malaria, developing an effective vaccine to prevent malaria onset is crucial.

### *6.1 Malaria Vaccine Components*

Malaria life cycle is quite complex, and as such, designing an effective vaccine remains an arduous task. Malaria vaccines can be subdivided into three distinct groups that target specific antigens: transmission-blocking vaccines (TBV), pre-erythrocytic vaccine, and blood-stage vaccine (Duffy and Patrick Gorres 2000).

**Table 1** *Toxoplasma gondii* virus-like particle vaccines in animal models

| Type | Vaccine antigen | Parasite burden reduction (%) | Survival (%) | References |
|---|---|---|---|---|
| Type I<br>GT1 strain | ROP18,<br>MIC18 | ROP18 (>30%)<br>MIC8 (>50%)<br>ROP18 + MIC8 (>60%) | 10 days<br>12 days<br>17 days | Lee et al. (2018a) |
| | IMC+ROP18 +MIC | Multi-antigen VLP (>80%)<br>Combination VLP (>50%) | 20% survival<br>0% survival | Lee et al. (2018b) |
| Type I<br>RH strain | MIC8 | IN route (>40%)<br>IM route (>10%) | IN route (100%)<br>IM route (60%) | Lee et al. (2017b) |
| | Chimeric VLP | – | 20 days | Guo et al. (2019) |
| Type II ME49 strain | IMC | IMC (>60%) | 100% | Lee et al. (2016a, b) |
| | AMA1 | AMA1 (> 50%) | 60% | Kim et al. (2020) |
| | ROP4,<br>ROP13 | ROP4 (>50%)<br>ROP13 (>20%)<br>ROP4 + ROP13 (>80%)<br>ROP (4 + 13) (>80%) | 100% | Kang et al. (2019, 2020) |
| | ROP18,<br>MIC8 | ROP18 (>60%)<br>MIC8 (>60%)<br>ROP18 + MIC8 (>80%) | – | Lee et al. (2018a) |

A hyphen denotes that the corresponding data was not reported in the referenced studies

In the TBV, the antigens currently being investigated are the Pfs25, Pfs230, Pfs48/45, and Pvs230. Pre-erythrocytic vaccines are generally designed based on the circumsporozoite protein (CSP), but there are vaccines targeting the sporozoites such as the *P. falciparum* sporozoites (PfSPZ). Finally, the blood-stage antigens are the most numerous of the three, which includes but not limited to apical membrane antigen 1 (AMA1), the merozoite surface proteins (MSP), erythrocyte binding antigen 175 (EBA-175), *P. falciparum* reticulocyte-binding protein homolog 5 (PfRh5), *P. falciparum* schizont egress antigen 1 (PfSEA1), *P. falciparum* glutamic-acid-rich protein (PfGARP), and many others.

**Table 2** Malaria virus-like particle vaccines in mice

| Source | Vaccine antigen | Immunity | References |
|---|---|---|---|
| *P. falciparum* | Pfs25 | • Reduced oocyst counts per mosquito<br>• Enhanced TRA<br>• Bolstered Pfs25 antigen-specific antibody response<br>• Strong transmission-blocking activity | Wetzel et al. (2019), Marini et al. (2019), Jones et al. (2013), Brune et al. (2016), Brune et al. (2017), Leneghan et al. (2017), Thrane et al. (2016) |
| | Pfs230 | • Elicited high level of TRA | Wetzel et al. (2019), Singh et al. (2019) |
| | Pfs47 | • Potent antibody response and strong TRA | Yenkoidiok-Douti et al. (2019) |
| | Pfs28 | • High anti-Pfs28 antibody response | Brune et al. (2017) |
| | CSP | • High antibody titer<br>• Diminished liver parasite load<br>• Protection against blood-stage malaria | Urakami et al. (2017), Janitzek et al. (2016), Kingston et al. (2019), Whitacre et al. (2015) |
| | Pfs48/45 | • High antibody titer and TRA | Singh et al. (2017, 2019) |
| | VAR2CSA | • High antibody titer | Thrane et al. (2016) |
| | EMP1 | • Strong IgG response elicited | Harmsen et al. (2020) |
| *P. vivax* | CSP | • Potent antibody induction<br>• Highly protective | Almeida et al. (2014), Andersson et al. (2017), Salman et al. (2017), Atcheson et al. (2018) |
| | CelTOS | • Modest protective efficacy | Alves et al. (2017) |
| | TRAP | • Sterile protection induced by Rv21 VLP vaccine | Atcheson et al. (2018) |
| *P. berghei* | AMA1 | • Prolonged survival and significantly reduced parasitemia | Lee et al. (2019b) |
| | MSP-8 | • Parasitemia reduction | Lee et al. (2020a) |
| | MSP-9 | • Prolonged survival and significantly reduced parasitemia | Lee et al. (2020b) |
| | TRAP | • Strong cellular and humoral immunity<br>Sterile protection | Cabral-Miranda et al. (2017, 2018) |
| *P. yoelii* | CSP | • Robust antibody and $CD8^+$ T cell responses | Pattinson et al. (2019) |

### 6.2 Experimental Malaria VLP Vaccines Using Animal Models

Numerous VLP-based malaria vaccines have been generated to date targeting various stages of the disease (Table 2). A chimeric VLP expressing the antigens of *P. falciparum* fused to the surface protein of the duck hepatitis B virus has been proposed as a potential TBV (Wetzel et al. 2019). Prime immunization using *P. falciparum* P47 antigen-expressing VLP vaccine conjugated to an *Acinetobacter* bacteriophage followed by boost immunization with the Pfs47 monomer elicited up to 98% transmission reducing activity (TRA) in mice (Yenkoidiok-Douti et al. 2019).

Another antigen from *P. falciparum* was used to generate a VLP vaccine expressed on top of the HBV surface antigen and its immunization in mice demonstrated high level of antibody response and TRA (Marini et al. 2019). A chimeric VLP vaccine expressing *P. falciparum* Pfs25 fused to Alfalfa mosaic virus coat protein completely blocked parasitic transmission over the span of six months (Jones et al. 2013). Immunizing mice with malarial antigens such as the Pfs25 or complex lysine and cysteine-rich inter-domain region (CIDR) expressed using SpyCatcher:SpyTag plug-and-display technology induced a strong antibody response with only a single immunization (Brune et al. 2016). Multimerizing the Pfs25 and Pfs28 protein with the SpyCatcher-IMX313-SnoopCatcher enhanced the murine antibody response to both antigens nearly 100-fold with only a single immunization (Brune et al. 2017). Several VLP display technologies have been assessed using the Pfs25 vaccines. Compared to Pfs25 protein genetically fused to the IMX313 heptamer, antibody inductions were vastly enhanced when the Pfs25 proteins were expressed using the SpyCatcher-AP205 VLPs or chemically conjugated to Qβ phage. The sheer amounts of antibodies induced were found to be the highest from Qβ-VLPs, whereas the highest quality of Pfs25-specific antibodies was elicited from SpyCatcher-AP205 VLP immunization (Leneghan et al. 2017).

*P. falciparum* CSP expressed on the envelope protein of the chikungunya virus induced a strong immune response upon challenge infection with the malaria sporozoites (Urakami et al. 2017). Although a challenge infection was not performed, strongly enhanced antibody responses were also observed from immunization with a VLP vaccine expressing the full-length CSP of *P. vivax* (Almeida et al. 2014; Andersson et al. 2017). Anchoring the VAR2CSA ID1-DBL2X-ID2a domain to the TM-CT of influenza A virus HA enhanced the antibody responses as well as parasite inhibition (Andersson et al. 2017). The Pfs48/45 VLP vaccines supplemented with adjuvants were confirmed to inhibit *P. falciparum* transmission (Singh et al. 2017,2019). Sander and his research group developed several VLP vaccines that induced long-lasting antibody responses in mice, which was maintained at high levels for months (Janitzek et al. 2016; Thrane et al. 2016). VLP vaccine designed using the Asp-Ala-Asp-Pro (NANP) repeat epitopes of the *P. falciparum* CSP has been proposed as a potential optimization strategy since the number of NANP was correlated with the degree of complement fixation to CSP

(Kingston et al. 2019). Conjugating the *P. falciparum* CIDRα1 domain of erythrocyte membrane protein 1 induced antibody responses, though more research needs to be conducted for its future application (Harmsen et al. 2020).

*P. vivax* is another species capable of causing human malaria. Vaccines designed using the *P. vivax* cell-traversal protein for ookinetes and sporozoites (CelTOS) have demonstrated interesting results. Priming with adenoviral vaccine followed by VLP or recombinant protein boost immunization conferred modest protection in mice (Alves et al. 2017). VLP vaccine expressing the CSP of *P. vivax* conferred 100% sterile protection against transgenic *P. berghei* challenge and is currently under further development for clinical evaluation (Salman et al. 2017). Similar results were observed from VLPs expressing the T and B cell epitopes of *P. falciparum* and *P. vivax* CSP epitopes (Whitacre et al. 2015). Synergistic effect of combinatorial vaccinations using CSP-expressing Rv21 VLP and viral vectored antigens expressing CSP and thrombospondin-related adhesive protein (TRAP) improved the protective efficacy of the vaccine compared to the vaccine used alone, which could also be increased through adjuvant use (Atcheson et al. 2018). A chimeric VLP vaccine expressing the *P. yoelii* CSP T cell or B cell epitopes induced $CD8^+$ T cell responses and antibody responses (Pattinson et al. 2019).

In stark contrast to *P. vivax* or *P. falciparum*, studies investigating the efficacy of rodent malaria VLP vaccines are extremely scarce. Several VLPs expressing the AMA-1, MSP-8, and MSP-9 of *P. berghei* elicited potent protection and prolonged the survival times, but failed to confer 100% protection in mice (Lee et al. 2019b, 2020a, b).

A VLP vaccine expressing the *P. vivax* thrombospondin-related adhesive protein (TRAP) adjuvanted with microcrystalline tyrosine (MCT) conferred partial protection against rodent malaria *P. berghei* (Cabral-Miranda et al. 2017). Recently, a *P. falciparum* thrombospondin-related adhesive protein (TRAP)-expressing VLP vaccine has been generated using the cucumber mosaic virus that has been chemically fused to the tetanus toxin T cell epitope. This VLP vaccine, when adjuvanted with the dioleoylphosphatidylserine (DOPS), induced stronger humoral and cellular immune responses than those induced by alum-adjuvanted VLPs upon challenge infection with transgenic *P. berghei* (Cabral-Miranda et al. 2018).

## *6.3 Clinical Malaria VLP Vaccines in Trial*

Several clinical trials have been conducted using various VLP-based malaria vaccines. A Pfs25 VLP vaccine produced from the plant *Nicotiana benthamiana* has recently underwent phase 1 study. Although substantial Pfs25-specific antibody responses were generated, the TRA was determined to be suboptimal (Chichester et al. 2018). A VLP vaccine expressing the PfRH5 constructed using the *Drosophila melanogaster* Schneider 2 (S2) cell line was confirmed to be highly immunogenic in mice and induced functional growth inhibitory antibodies against *P. falciparum* in vitro. As such, this RH5.1 VLP vaccine adjuvanted with the $AS01_B$ has been approved for clinical trial phase I/IIa in the United Kingdom (Jin et al. 2018). Phase III clinical trial results for the malaria vaccine RTS,S/AS01 have reported that the vaccines initially protected the patients against malaria, but its protective efficacy waned over time (Olotu et al. 2016). With the disappointing efficacy demonstrated by RTS,S, developing an alternative vaccine using different strategies has become a necessity. Recently, at the University of Oxford's Jenner Institute, a RTS,S-like vaccine was designed which was labeled as R21 that were more immunogenic than its predecessor RTS,S. This vaccine was reported to be undergoing several clinical trials in the United Kingdom and West Africa (Collins et al. 2017).

## 7 Conclusion

Developing efficacious vaccines for diseases is extremely difficult and more efforts are needed to eradicate some of the infectious diseases of global importance. Advances in biotechnology have enabled construction of VLP vaccines and the efficacies of these vaccines appear to be promising. With the recent emergence of the novel coronavirus pandemic affecting the entire world, rapid production of vaccines with outstanding efficacy such as VLPs has become a global necessity.

**Funding Information** This work was supported by grants from the Ministry of Health and Welfare, Republic of Korea (HV20C0085, HV20C0142), and the National Research Foundation of Korea (NRF) (2018R1A6A1A03025124).

## References

Akerlind-Stopner B, Hu A, Mufson MA, Utter G, Norrby E (1993) Antibody responses of children to the C-terminal peptide of the SH protein of respiratory syncytial virus and the immunological characterization of this protein. J Med Virol 40(2):112–120

Almeida AP, Dias MO, Vieira Cde A, Chávez-Olórtegui C, Gazzineli RT, Rodrigues MM, Fujiwara RT, Bruna-Romero O (2014) Long-lasting humoral and cellular immune responses elicited by immunization with recombinant chimeras of the plasmodium vivax circumsporozoite protein. Vaccine 32(19):2181–2187

Alves E, Salman AM, Leoratti F, Lopez-Camacho C, Viveros-Sandoval ME, Lall A, El-Turabi A, Bachmann MF, Hill AV, Janse CJ, Khan SM et al (2017) Evaluation of plasmodium vivax cell-traversal protein for ookinetes and sporozoites as a preerythrocytic P. Vivax vaccine. Clin Vaccine Immunol 24(4):e00501–16

Andersson AC, Resende M, Salanti A, Nielsen MA, Holst PJ (2017) Novel adenovirus encoded virus-like particles displaying the placental malaria associated var2csa antigen. Vaccine 35 (8):1140–1147

Atcheson E, Bauza K, Salman AM, Alves E, Blight J, Viveros-Sandoval ME, Janse CJ, Khan SM, Hill AVS, Reyes-Sandoval A (2018) Tailoring a plasmodium vivax vaccine to enhance efficacy through a combination of a csp virus-like particle and trap viral vectors. Infect Immun 24(4): e00114–18

Bachmann MF, Jennings GT (2010) Vaccine delivery: a matter of size, geometry, kinetics and molecular patterns. Nat Rev Immunol 10(11):787–796

Bachmann MF, Rohrer UH, Kündig TM, Bürki K, Hengartner H, Zinkernagel RM (1993) The influence of antigen organization on b cell responsiveness. Science 262(5138):1448–1451

Berlanda Scorza F, Tsvetnitsky V, Donnelly JJ (2016) Universal influenza vaccines: shifting to better vaccines. Vaccine 34(26):2926–2933

Black MW, Boothroyd JC (2000) Lytic cycle of toxoplasma gondii. Microbiol Mol Biol Rev 64 (3):607–623

Blanco JCG, Pletneva LM, McGinnes-Cullen L, Otoa RO, Patel MC, Fernando LR, Boukhvalova MS, Morrison TG (2018) Efficacy of a respiratory syncytial virus vaccine candidate in a maternal immunization model. Nat Commun 9(1):1904

Bragard C, Duncan GH, Wesley SV, Naidu RA, Mayo MA (2000) Virus-like particles assemble in plants and bacteria expressing the coat protein gene of Indian peanut clump virus. J Gen Virol 81(Pt 1):267–272

Brune KD, Howarth M (2018) New routes and opportunities for modular construction of particulate vaccines: stick, click, and glue. Front Immunol 9:1432

Brune KD, Leneghan DB, Brian IJ, Ishizuka AS, Bachmann MF, Draper SJ, Biswas S, Howarth M (2016) Plug-and-display: decoration of virus-like particles via isopeptide bonds for modular immunization. Sci Rep 6:19234

Brune KD, Buldun CM, Li Y, Taylor IJ, Brod F, Biswas S, Howarth M (2017) Dual plug-and-display synthetic assembly using orthogonal reactive proteins for twin antigen immunization. Bioconjug Chem 28(5):1544–1551

Cabral-Miranda G, Heath MD, Mohsen MO, Gomes AC, Engeroff P, Flaxman A, Leoratti FMS, El-Turabi A, Reyes-Sandoval A, Skinner MA, Kramer MF et al (2015) Virus-like particle (VLP) plus microcrystalline tyrosine (mct) adjuvants enhance vaccine efficacy improving t and b cell immunogenicity and protection against plasmodium berghei/vivax. Vaccines (Basel) 5(2)

Cabral-Miranda G, Salman AM, Mohsen MO, Storni FL, Roesti ES, Skinner MA, Heath MD, Kramer MF, Khan SM, Janse CJ, Hill AVS et al (2018) Dops adjuvant confers enhanced protection against malaria for VLP-trap based vaccines. Diseases 6(4):107

Cane PA (2001) Molecular epidemiology of respiratory syncytial virus. Rev Med Virol 11(2):103–116

Chackerian B (2007) Virus-like particles: flexible platforms for vaccine development. Expert Rev Vaccines 6(3):381–390

Charlton Hume HK, Lua LHL (2017) Platform technologies for modern vaccine manufacturing. Vaccine 35(35 Pt A):4480–4485

Chichester JA, Green BJ, Jones RM, Shoji Y, Miura K, Long CA, Lee CK, Ockenhouse CF, Morin MJ, Streatfield SJ, Yusibov V (2018) Safety and immunogenicity of a plant-produced Pfs25 virus-like particle as a transmission blocking vaccine against malaria: a phase 1 dose-escalation study in healthy adults. Vaccine 36(39):5865–5871

Chung KY, Coyle EM, Jani D, King LR, Bhardwaj R, Fries L, Smith G, Glenn G, Golding H, Khurana S (2015) Iscomatrix™ adjuvant promotes epitope spreading and antibody affinity maturation of influenza a H7N9 virus like particle vaccine that correlate with virus neutralization in humans. Vaccine 33(32):3953–3962

Cimica V, Boigard H, Bhatia B, Fallon JT, Alimova A, Gottlieb P, Galarza JM (2016) Novel respiratory syncytial virus-like particle vaccine composed of the postfusion and prefusion conformations of the f glycoprotein. Clin Vaccine Immunol 23(6):451–459

Collins PL, Fearns R, Graham BS (2013) Respiratory syncytial virus: virology, reverse genetics, and pathogenesis of disease. Curr Top Microbiol Immunol 372:3–38

Collins KA, Snaith R, Cottingham MG, Gilbert SC, Hill AVS (2017) Enhancing protective immunity to malaria with a highly immunogenic virus-like particle vaccine. Sci Rep 7:46621

Cox MM, Hollister JR (2009) Flublok, a next generation influenza vaccine manufactured in insect cells. Biologicals 37(3):182–189

Cullen LM, Boukhvalova MS, Blanco JCG, Morrison TG (2020) Comparisons of antibody populations in different pre-fusion f VLP-immunized cotton rat dams and their offspring. Vaccines (Basel) 8(1):133

Danwang C, Bigna JJ, Nzalie RNT, Robert A (2020) Epidemiology of clinical congenital and neonatal malaria in endemic settings: A systematic review and meta-analysis. Malar J 19 (1):312

DiMiceli L, Pool V, Kelso JM, Shadomy SV, Iskander J (2006) Vaccination of yeast sensitive individuals: review of safety data in the US vaccine adverse event reporting system (vaers). Vaccine 24(6):703–707

Donaldson B, Al-Barwani F, Young V, Scullion S, Ward V, Young S (2015) Virus-like particles, a versatile subunit vaccine platform. In: Foged C, Rades T, Perrie Y, Hook S (eds) Subunit vaccine delivery. New York, NY, Springer, New York, pp 159–180

Dubey JP (2009) Toxoplasmosis in sheep–the last 20 years. Vet Parasitol 163(1–2):1–14

Duffy PE, Patrick Gorres J (2000) Malaria vaccines since 2000: Progress, priorities, products. NPJ Vaccines 5:48

Effio CL, Hubbuch J (2015) Next generation vaccines and vectors: designing downstream processes for recombinant protein-based virus-like particles. Biotechnol J 10(5):715–727

Efstathiou C, Abidi SH, Harker J, Stevenson NJ (2020) Revisiting respiratory syncytial virus's interaction with host immunity, towards novel therapeutics. Cell Mol Life Sci 1–14

Eichinger KM, Kosanovich JL, Gidwani SV, Zomback A, Lipp MA, Perkins TN, Oury TD, Petrovsky N, Marshall CP, Yondola MA, Empey KM (2020) Prefusion rsv f immunization elicits th2-mediated lung pathology in mice when formulated with a th2 (but not a th1/th2-balanced) adjuvant despite complete viral protection. Front Immunol 11:1673

Fehr T, Skrastina D, Pumpens P, Zinkernagel RM (1998) T cell-independent type i antibody response against b cell epitopes expressed repetitively on recombinant virus particles. Proc Natl Acad Sci U S A 95(16):9477–9481

Frey SJ, Varner C, Arsiwala A, Currier MG, Moore ML, Kane RS (2020) The design of vaccines based on the shielding of antigenic site Ú of a respiratory syncytial virus fusion protein immunogen. Adv Healthc Mater e2000714

Gause KT, Wheatley AK, Cui J, Yan Y, Kent SJ, Caruso F (2017) Immunological principles guiding the rational design of particles for vaccine delivery. ACS Nano 11(1):54–68

Giles BM, Ross TM (2011) A computationally optimized broadly reactive antigen (cobra) based H5N1 VLP vaccine elicits broadly reactive antibodies in mice and ferrets. Vaccine 29 (16):3043–3054

Gomes AC, Mohsen M, Bachmann MF (2017) Harnessing nanoparticles for immunomodulation and vaccines. Vaccines (Basel) 5(1):6

Guo J, Zhou A, Sun X, Sha W, Ai K, Pan G, Zhou C, Zhou H, Cong H, He S (2009) Immunogenicity of a virus-like-particle vaccine containing multiple antigenic epitopes of toxoplasma gondii against acute and chronic toxoplasmosis in mice. Front Immunol 10:592

Ha B, Yang JE, Chen X, Jadhao SJ, Wright ER, Anderson LJ (2020) Two rsv platforms for g, f, or g+f proteins vlps. Viruses 12(9):906

Harmsen C, Turner L, Thrane S, Sander AF, Theander TG, Lavstsen T (2020) Immunization with virus-like particles conjugated to cidrα1 domain of plasmodium falciparum erythrocyte membrane protein 1 induces inhibitory antibodies. Malar J 19(1):132

Hervas-Stubbs S, Rueda P, Lopez L, Leclerc C (2007) Insect baculoviruses strongly potentiate adaptive immune responses by inducing type I IFN. J Immunol 178(4):2361–2369

Hodgins B, Pillet S, Landry N, Ward BJ (2019) A plant-derived vlp influenza vaccine elicits a balanced immune response even in very old mice with co-morbidities. PLoS ONE 14(1): e0210009

Huang Z, Santi L, LePore K, Kilbourne J, Arntzen CJ, Mason HS (2006) Rapid, high-level production of hepatitis b core antigen in plant leaf and its immunogenicity in mice. Vaccine 24 (14):2506–2513

Hwang HS, Kwon YM, Lee JS, Yoo SE, Lee YN, Ko EJ, Kim MC, Cho MK, Lee YT, Jung YJ, Lee JY et al (2014) Co-immunization with virus-like particle and DNA vaccines induces protection against respiratory syncytial virus infection and bronchiolitis. Antiviral Res 110:115–123

Hwang HS, Lee YT, Kim KH, Park S, Kwon YM, Lee Y, Ko EJ, Jung YJ, Lee JS, Kim YJ, Lee YN et al (2016) Combined virus-like particle and fusion protein-encoding DNA vaccination of cotton rats induces protection against respiratory syncytial virus without causing vaccine-enhanced disease. Virology 494:215–224

Hwang HS, Kim KH, Lee Y, Lee YT, Ko EJ, Park S, Lee JS, Lee BC, Kwon YM, Moore ML, Kang SM (2017a) Virus-like particle vaccines containing f or f and g proteins confer protection against respiratory syncytial virus without pulmonary inflammation in cotton rats. Hum Vaccin Immunother 13(5):1031–1039

Hwang HS, Lee YT, Kim KH, Ko EJ, Lee Y, Kwon YM, Kang SM (2017b) Virus-like particle vaccine primes immune responses preventing inactivated-virus vaccine-enhanced disease against respiratory syncytial virus. Virology 511:142–151

Jang YH, Seong BL (2020) Call for a paradigm shift in the design of universal influenza vaccines by harnessing multiple correlates of protection. Expert Opin Drug Discov 15(12):1441–1455

Janitzek CM, Matondo S, Thrane S, Nielsen MA, Kavishe R, Mwakalinga SB, Theander TG, Salanti A, Sander AF (2016) Bacterial superglue generates a full-length circumsporozoite protein virus-like particle vaccine capable of inducing high and durable antibody responses. Malar J 15(1):545

Jennings GT, Bachmann MF (2008) The coming of age of virus-like particle vaccines. Biol Chem 389(5):521–536

Jiang W, Wang S, Chen H, Ren H, Huang X, Wang G, Chen Z, Chen L, Chen Z, Zhou P (2017) A bivalent heterologous DNA virus-like-particle prime-boost vaccine elicits broad protection against both group 1 and 2 influenza a viruses. J Virol 91(9):e02052–16

Jin J, Tarrant RD, Bolam EJ, Angell-Manning P, Soegaard M, Pattinson DJ, Dulal P, Silk SE, Marshall JM, Dabbs RA, Nugent FL et al (2018) Production, quality control, stability, and potency of cGMP-produced plasmodium falciparum RH5.1 protein vaccine expressed in drosophila s2 cells. NPJ Vaccines 3:32

Jones RM, Chichester JA, Mett V, Jaje J, Tottey S, Manceva S, Casta LJ, Gibbs SK, Musiychuk K, Shamloul M, Norikane J et al (2013) A plant-produced Pfs25 VLP malaria vaccine candidate induces persistent transmission blocking antibodies against plasmodium falciparum in immunized mice. PLoS ONE 8(11):e79538

Kang HJ, Baker EN (2011) Intramolecular isopeptide bonds: protein crosslinks built for stress? Trends Biochem Sci 36(4):229–237

Kang HJ, Lee SH, Kim MJ, Chu KB, Lee DH, Chopra M, Choi HJ, Park H, Jin H, Quan FS (2019) Influenza virus-like particles presenting both toxoplasma gondii rop4 and rop13 enhance protection against T. Gondii infection. Pharmaceutics 11(7):342

Kang HJ, Chu KB, Lee SH, Kim MJ, Park H, Jin H, Moon EK, Quan FS (2020) Toxoplasma gondii virus-like particle vaccination alleviates inflammatory response in the brain upon T. gondii infection. Parasite Immunol 42(6):e12716

Khurana S, Wu J, Verma N, Verma S, Raghunandan R, Manischewitz J, King LR, Kpamegan E, Pincus S, Smith G, Glenn G et al (2011) H5n1 virus-like particle vaccine elicits cross-reactive neutralizing antibodies that preferentially bind to the oligomeric form of influenza virus hemagglutinin in humans. J Virol 85(21):10945–10954

Khuu D, Eberhard ML, Bristow BN, Javanbakht M, Ash LR, Shafir SC, Sorvillo FJ (2019) Economic impact of malaria-related hospitalizations in the united states, 2000–2014. J Infect Public Health 12(3):424–433

Kim MC, Song JM, O E, Kwon YM, Lee YJ, Compans RW, Kang SM (2013) Virus-like particles containing multiple m2 extracellular domains confer improved cross-protection against various subtypes of influenza virus. Mol Ther 21(2):485–492

Kim MC, Lee YN, Ko EJ, Lee JS, Kwon YM, Hwang HS, Song JM, Song BM, Lee YJ, Choi JG, Kang HM et al (2014) Supplementation of influenza split vaccines with conserved m2 ectodomains overcomes strain specificity and provides long-term cross protection. Mol Ther 22 (7):1364–1374

Kim KH, Lee YT, Hwang HS, Kwon YM, Kim MC, Ko EJ, Lee JS, Lee Y, Kang SM (2015) Virus-like particle vaccine containing the f protein of respiratory syncytial virus confers protection without pulmonary disease by modulating specific subsets of dendritic cells and effector t cells. J Virol 89(22):11692–11705

Kim YJ, Lee YT, Kim MC, Lee YN, Kim KH, Ko EJ, Song JM, Kang SM (2017) Cross-protective efficacy of influenza virus m2e containing virus-like particles is superior to hemagglutinin vaccines and variable depending on the genetic backgrounds of mice. Front Immunol 8:1730

Kim AR, Lee DH, Lee SH, Rubino I, Choi HJ, Quan FS (2018) Protection induced by virus-like particle vaccine containing tandem repeat gene of respiratory syncytial virus g protein. PLoS ONE 13(1):e0191277

Kim MJ, Lee SH, Kang HJ, Chu KB, Park H, Jin H, Moon EK, Kim SS, Quan FS (2020) Virus-like particle vaccine displaying toxoplasma gondii apical membrane antigen 1 induces protection against T. Gondii ME49 infection in mice. Microb Pathog 142:104090

Kingston NJ, Kurtovic L, Walsh R, Joe C, Lovrecz G, Locarnini S, Beeson JG, Netter HJ (2019) Hepatitis B virus-like particles expressing plasmodium falciparum epitopes induce complement-fixing antibodies against the circumsporozoite protein. Vaccine 37(12):1674–1684

Ko EJ, Kwon YM, Lee JS, Hwang HS, Yoo SE, Lee YN, Lee YT, Kim MC, Cho MK, Lee YR, Quan FS et al (2015) Virus-like nanoparticle and DNA vaccination confers protection against respiratory syncytial virus by modulating innate and adaptive immune cells. Nanomedicine 11 (1):99–108

Krammer F, Palese P (2015) Advances in the development of influenza virus vaccines. Nat Rev Drug Discov 14(3):167–182

Kramer F, García-Sastre A, Palese P (2018) Is it possible to develop a "universal" influenza virus vaccine? Potential target antigens and critical aspects for a universal influenza vaccine. Cold Spring Harb Perspect Biol 10(7):a028845

Krietsch Boerner L (2020) The flu shot and the egg. ACS Cent Sci 6(2):89–92

Kushnir N, Streatfield SJ, Yusibov V (2012) Virus-like particles as a highly efficient vaccine platform: diversity of targets and production systems and advances in clinical development. Vaccine 31(1):58–83

Landry N, Ward BJ, Trépanier S, Montomoli E, Dargis M, Lapini G, Vézina LP (2010) Preclinical and clinical development of plant-made virus-like particle vaccine against avian h5n1 influenza. PLoS ONE 5(12):e15559

Landry N, Pillet S, Favre D, Poulin JF, Trépanier S, Yassine-Diab B, Ward BJ (2014) Influenza virus-like particle vaccines made in *Nicotiana benthamiana* elicit durable, poly-functional and cross-reactive t cell responses to influenza ha antigens. Clin Immunol 154(2):164–177

Lee S, Quan FS, Kwon Y, Sakamoto K, Kang SM, Compans RW, Moore ML (2014) Additive protection induced by mixed virus-like particles presenting respiratory syncytial virus fusion or attachment glycoproteins. Antiviral Res 111:129–135

Lee DH, Lee SH, Kim AR, Quan FS (2016a) Virus-like nanoparticle vaccine confers protection against toxoplasma gondii. PLoS ONE 11(8):e0161231

Lee DH, Kim AR, Lee SH, Quan FS (2016b) Cross-protection induced by toxoplasma gondii virus-like particle vaccine upon intraperitoneal route challenge. Acta Trop 164:77–83

Lee Y, Lee YT, Ko EJ, Kim KH, Hwang HS, Park S, Kwon YM, Kang SM (2017a) Soluble f proteins exacerbate pulmonary histopathology after vaccination upon respiratory syncytial virus challenge but not when presented on virus-like particles. Hum Vaccin Immunother 13 (11):2594–2605

Lee SH, Kim AR, Lee DH, Rubino I, Choi HJ, Quan FS (2017b) Protection induced by virus-like particles containing toxoplasma gondii microneme protein 8 against highly virulent RH strain of toxoplasma gondii infection. PLoS ONE 12(4):e0175644

Lee SH, Kang HJ, Lee DH, Kang SM, Quan FS (2018a) Virus-like particle vaccines expressing toxoplasma gondii rhoptry protein 18 and microneme protein 8 provide enhanced protection. Vaccine 36(38):5692–5700

Lee SH, Kang HJ, Lee DH, Quan FS (2018b) Protective immunity induced by incorporating multiple antigenic proteins of toxoplasma gondii into influenza virus-like particles. Front Immunol 9:3073

Lee DH, Chu KB, Kang HJ, Lee SH, Chopra M, Choi HJ, Moon EK, Inn KS, Quan FS (2019a) Protection induced by malaria virus-like particles containing codon-optimized AMA-1 of plasmodium berghei. Malar J 18(1):394

Lee YT, Kim KH, Ko EJ, Kim MC, Lee YN, Hwang HS, Lee Y, Jung YJ, Kim YJ, Santos J, Perez DR et al (2019b) Enhancing the cross protective efficacy of live attenuated influenza virus vaccine by supplemented vaccination with m2 ectodomain virus-like particles. Virology 529:111–121

Lee SH, Chu KB, Kang HJ, Basak S, Kim MJ, Park H, Jin H, Moon EK, Quan FS (2020a) Virus-like particles expressing plasmodium berghei msp-8 induce protection against P. Berghei infection. Parasite Immunol 42(11):e12781

Lee SH, Kang HJ, Chu KB, Basak S, Lee DH, Moon EK, Quan FS (2020b) Protective immunity induced by virus-like particle containing merozoite surface protein 9 of plasmodium berghei. Vaccines (Basel) 8(3):428

Leneghan DB, Miura K, Taylor IJ, Li Y, Jin J, Brune KD, Bachmann MF, Howarth M, Long CA, Biswas S (2017) Nanoassembly routes stimulate conflicting antibody quantity and quality for transmission-blocking malaria vaccines. Sci Rep 7(1):3811

Li Y, Zhou H (2018) Moving towards improved vaccines for toxoplasma gondii. Expert Opin Biol Ther 18(3):273–280

Lindsay BJ, Bonar MM, Costas-Cancelas IN, Hunt K, Makarkov AI, Chierzi S, Krawczyk CM, Landry N, Ward BJ, Rouiller I (2018) Morphological characterization of a plant-made virus-like particle vaccine bearing influenza virus hemagglutinins by electron microscopy. Vaccine 36(16):2147–2154

Ling HY, Edwards AM, Gantier MP, Deboer KD, Neale AD, Hamill JD, Walmsley AM (2012) An interspecific Nicotiana hybrid as a useful and cost-effective platform for production of animal vaccines. PLoS ONE 7(4):e35688

Liu F, Wu X, Li L, Liu Z, Wang Z (2013) Use of baculovirus expression system for generation of virus-like particles: successes and challenges. Protein Expr Purif 90(2):104–116

Lomonossoff GP, D'Aoust MA (2016) Plant-produced biopharmaceuticals: a case of technical developments driving clinical deployment. Science 353(6305):1237–1240

López-Macías C, Ferat-Osorio E, Tenorio-Calvo A, Isibasi A, Talavera J, Arteaga-Ruiz O, Arriaga-Pizano L, Hickman SP, Allende M, Lenhard K, Pincus S et al (2011) Safety and immunogenicity of a virus-like particle pandemic influenza a (H1N1) 2009 vaccine in a blinded, randomized, placebo-controlled trial of adults in Mexico. Vaccine 29(44):7826–7834

Low JG, Lee LS, Ooi EE, Ethirajulu K, Yeo P, Matter A, Connolly JE, Skibinski DA, Saudan P, Bachmann M, Hanson BJ et al (2014) Safety and immunogenicity of a virus-like particle pandemic influenza a (H1N1) 2009 vaccine: results from a double-blinded, randomized phase i clinical trial in healthy Asian volunteers. Vaccine 32(39):5041–5048

Lua LH, Connors NK, Sainsbury F, Chuan YP, Wibowo N, Middelberg AP (2014) Bioengineering virus-like particles as vaccines. Biotechnol Bioeng 111(3):425–440

Luo WY, Lin SY, Lo KW, Lu CH, Hung CL, Chen CY, Chang CC, Hu YC (2013) Adaptive immune responses elicited by baculovirus and impacts on subsequent transgene expression in vivo. J Virol 87(9):4965–4973

Madhi SA, Polack FP, Piedra PA, Munoz FM, Trenholme AA, Simões EAF, Swamy GK, Agrawal S, Ahmed K, August A, Baqui AH et al (2020) Respiratory syncytial virus vaccination during pregnancy and effects in infants. N Engl J Med 383(5):426–439

Malaria: World Health Organization, Geneva (2020). https://www.who.int/news-room/fact-sheets/detail/malaria

Manolova V, Flace A, Bauer M, Schwarz K, Saudan P, Bachmann MF (2008) Nanoparticles target distinct dendritic cell populations according to their size. Eur J Immunol 38(5):1404–1413

Marini A, Zhou Y, Li Y, Taylor IJ, Leneghan DB, Jin J, Zaric M, Mekhaiel D, Long CA, Miura K, Biswas S (2019) A universal plug-and-display vaccine carrier based on HBsAg VLP to maximize effective antibody response. Front Immunol 10:2931

Mason HS, Ball JM, Shi JJ, Jiang X, Estes MK, Arntzen CJ (1996) Expression of Norwalk virus capsid protein in transgenic tobacco and potato and its oral immunogenicity in mice. Proc Natl Acad Sci U S A 93(11):5335–5340

Mazur NI, Higgins D, Nunes MC, Melero JA, Langedijk AC, Horsley N, Buchholz UJ, Openshaw PJ, McLellan JS, Englund JA, Mejias A et al (2018) The respiratory syncytial virus vaccine landscape: lessons from the graveyard and promising candidates. Lancet Infect Dis 18 (10):e295–e311

McGinnes LW, Gravel KA, Finberg RW, Kurt-Jones EA, Massare MJ, Smith G, Schmidt MR, Morrison TG (2011) Assembly and immunological properties of newcastle disease virus-like particles containing the respiratory syncytial virus f and g proteins. J Virol 85(1):366–377

McGinnes Cullen L, Schmidt MR, Kenward SA, Woodland RT, Morrison TG (2015) Murine immune responses to virus-like particle-associated pre- and postfusion forms of the respiratory syncytial virus f protein. J Virol 89(13):6835–6847

McGinnes Cullen L, Schmidt MR, Morrison TG (2019) Effect of previous respiratory syncytial virus infection on murine immune responses to f and g protein-containing virus-like particles. J Virol 93(9):e00087–19

McLellan JS, Chen M, Leung S, Graepel KW, Du X, Yang Y, Zhou T, Baxa U, Yasuda E, Beaumont T, Kumar A et al (2013) Structure of RSV fusion glycoprotein trimer bound to a prefusion-specific neutralizing antibody. Science 340(6136):1113–1117

Mohsen MO, Zha L, Cabral-Miranda G, Bachmann MF (2017a) Major findings and recent advances in virus-like particle (vlp)-based vaccines. Semin Immunol 34:123–132

Mohsen MO, Gomes AC, Cabral-Miranda G, Krueger CC, Leoratti FM, Stein JV, Bachmann MF (2017b) Delivering adjuvants and antigens in separate nanoparticles eliminates the need of physical linkage for effective vaccination. J Control Release 251:92–100

Molinari NA, Ortega-Sanchez IR, Messonnier ML, Thompson WW, Wortley PM, Weintraub E, Bridges CB (2007) The annual impact of seasonal influenza in the US: measuring disease burden and costs. Vaccine 25(27):5086–5096

Montazeri M, Mikaeili Galeh T, Moosazadeh M, Sarvi S, Dodangeh S, Javidnia J, Sharif M, Daryani A (2020) The global serological prevalence of toxoplasma gondii in felids during the last five decades (1967–2017): a systematic review and meta-analysis. Parasit Vectors 13(1):82

Morón G, Rueda P, Casal I, Leclerc C (2002) Cd8alpha- cd11b+ dendritic cells present exogenous virus-like particles to $cd8^{+}$ t cells and subsequently express cd8alpha and cd205 molecules. J Exp Med 195(10):1233–1245

Moyle PM, Toth I (2013) Modern subunit vaccines: Development, components, and research opportunities. ChemMedChem 8(3):360–376

Murawski MR, McGinnes LW, Finberg RW, Kurt-Jones EA, Massare MJ, Smith G, Heaton PM, Fraire AE, Morrison TG (2010) Newcastle disease virus-like particles containing respiratory syncytial virus g protein induced protection in BALB/c mice, with no evidence of immunopathology. J Virol 84(2):1110–1123

Music N, Reber AJ, Kim MC, York IA, Kang SM (2016) Supplementation of H1N1pdm09 split vaccine with heterologous tandem repeat m2e5x virus-like particles confers improved cross-protection in ferrets. Vaccine 34(4):466–473

Ngwuta JO, Chen M, Modjarrad K, Joyce MG, Kanekiyo M, Kumar A, Yassine HM, Moin SM, Killikelly AM, Chuang GY, Druz A et al (2015) Prefusion f-specific antibodies determine the magnitude of RSV neutralizing activity in human sera. Sci Transl Med 7(309):309ra162

Olotu A, Fegan G, Wambua J, Nyangweso G, Leach A, Lievens M, Kaslow DC, Njuguna P, Marsh K, Bejon P (2016) Seven-year efficacy of RTS, S/AS01 malaria vaccine among young African children. N Engl J Med 374(26):2519–2529

Pappas G, Roussos N, Falagas ME (2009) Toxoplasmosis snapshots: global status of toxoplasma gondii seroprevalence and implications for pregnancy and congenital toxoplasmosis. Int J Parasitol 39(12):1385–1394

Park YC, Song JM (2017) Preparation and immunogenicity of influenza virus-like particles using nitrocellulose membrane filtration. Clin Exp Vaccine Res 6(1):61–66

Pattinson DJ, Apte SH, Wibowo N, Chuan YP, Rivera-Hernandez T, Groves PL, Lua LH, Middelberg APJ, Doolan DL (2019) Chimeric murine polyomavirus virus-like particles induce plasmodium antigen-specific cd8(+) t cell and antibody responses. Front Cell Infect Microbiol 9:215

Pillet S, Couillard J, Trépanier S, Poulin JF, Yassine-Diab B, Guy B, Ward BJ, Landry N (2019) Immunogenicity and safety of a quadrivalent plant-derived virus like particle influenza vaccine candidate-two randomized phase ii clinical trials in 18 to 49 and $\geq 50$ years old adults. PLoS ONE 14(6):e0216533

Plotkin S (2014) History of vaccination. Proc Natl Acad Sci U S A 111(34):12283–12287

Plotkin S, Robinson JM, Cunningham G, Iqbal R, Larsen S (2017) The complexity and cost of vaccine manufacturing—an overview. Vaccine 35(33):4064–4071

Puente-Massaguer E, Gòdia F, Lecina M (2020) Development of a non-viral platform for rapid virus-like particle production in sf9 cells. J Biotechnol 322:43–53

Qin J, Zhang Y, Shen X, Gong L, Peng O, Liu Y, Xue C, Cao Y (2018) H7 virus-like particles assembled by hemagglutinin containing H3N2 transmembrane domain and m1 induce broad homologous and heterologous protection in mice. Vaccine 36(33):5030–5036

Quan FS, Vunnava A, Compans RW, Kang SM (2010) Virus-like particle vaccine protects against 2009 h1n1 pandemic influenza virus in mice. PLoS ONE 5(2):e9161

Quan FS, Kim Y, Lee S, Yi H, Kang SM, Bozja J, Moore ML, Compans RW (2011) Virus like particle vaccine induces protection against respiratory syncytial virus infection in mice. J Infect Dis 204(7):987–995

Quan FS, Kim MC, Lee BJ, Song JM, Compans RW, Kang SM (2012) Influenza M1 VLPS containing neuraminidase induce heterosubtypic cross-protection. Virology 430(2):127–135

Quan FS, Basak S, Chu KB, Kim SS, Kang SM (2020) Progress in the development of virus-like particle vaccines against respiratory viruses. Expert Rev Vaccines 19(1):11–24

Raghunandan R, Lu H, Zhou B, Xabier MG, Massare MJ, Flyer DC, Fries LF, Smith GE, Glenn GM (2014) An insect cell derived respiratory syncytial virus (RSV) f nanoparticle vaccine induces antigenic site ii antibodies and protects against RSV challenge in cotton rats by active and passive immunization. Vaccine 32(48):6485–6492

Rezaei F, Sarvi S, Sharif M, Hejazi SH, Pagheh AS, Aghayan SA, Daryani A (2019) A systematic review of toxoplasma gondii antigens to find the best vaccine candidates for immunization. Microb Pathog 126:172–184

Rohovie MJ, Nagasawa M, Swartz JR (2017) Virus-like particles: next-generation nanoparticles for targeted therapeutic delivery. Bioeng Transl Med 2(1):43–57

Roldão A, Mellado MC, Castilho LR, Carrondo MJ, Alves PM (2010) Virus-like particles in vaccine development. Expert Rev Vaccines 9(10):1149–1176

Rueda P, Fominaya J, Langeveld JP, Bruschke C, Vela C, Casal JI (2000) Effect of different baculovirus inactivation procedures on the integrity and immunogenicity of porcine parvovirus-like particles. Vaccine 19(7–8):726–734

Ryu S, Cowling BJ (2020) Human influenza epidemiology. Cold Spring Harb Perspect Med

Salman AM, Montoya-Díaz E, West H, Lall A, Atcheson E, Lopez-Camacho C, Ramesar J, Bauza K, Collins KA, Brod F, Reis F et al (2017) Rational development of a protective p. Vivax vaccine evaluated with transgenic rodent parasite challenge models. Sci Rep 7:46482

Sarma N, Patouillard E, Cibulskis RE, Arcand JL (2019) The economic burden of malaria: revisiting the evidence. Am J Trop Med Hyg 101(6):1405–1415

Schepens B, Sedeyn K, Vande Ginste L, De Baets S, Schotsaert M, Roose K, Houspie L, Van Ranst M, Gilbert B, van Rooijen N, Fiers W et al (2014) Protection and mechanism of action of a novel human respiratory syncytial virus vaccine candidate based on the extracellular domain of small hydrophobic protein. EMBO Mol Med 6(11):1436–1454

Schwartzman LM, Cathcart AL, Pujanauski LM, Qi L, Kash JC, Taubenberger JK (2015) An intranasal virus-like particle vaccine broadly protects mice from multiple subtypes of influenza a virus. MBio 6(4):e01044

Shi T, McAllister DA, O'Brien KL, Simoes EAF, Madhi SA, Gessner BD, Polack FP, Balsells E, Acacio S, Aguayo C, Alassani I et al (2017) Global, regional, and national disease burden estimates of acute lower respiratory infections due to respiratory syncytial virus in young children in 2015: A systematic review and modelling study. Lancet 390(10098):946–958

Short KR, Richard M, Verhagen JH, van Riel D, Schrauwen EJ, van den Brand JM, Mänz B, Bodewes R, Herfst S (2015) One health, multiple challenges: the inter-species transmission of influenza a virus. One Health 1:1–13

Singh SK, Thrane S, Janitzek CM, Nielsen MA, Theander TG, Theisen M, Salanti A, Sander AF (2017) Improving the malaria transmission-blocking activity of a plasmodium falciparum 48/45 based vaccine antigen by SpyTag/SpyCatcher mediated virus-like display. Vaccine 35 (30):3726–3732

Singh SK, Thrane S, Chourasia BK, Teelen K, Graumans W, Stoter R, van Gemert GJ, van de Vegte-Bolmer MG, Nielsen MA, Salanti A, Sander AF et al (2019) Pfs230 and pfs48/45 fusion proteins elicit strong transmission-blocking antibody responses against plasmodium falciparum. Front Immunol 10:1256

Skibinski DAG, Jones LA, Zhu YO, Xue LW, Au B, Lee B, Naim ANM, Lee A, Kaliaperumal N, Low JGH, Lee LS et al (2018) Induction of human t-cell and cytokine responses following vaccination with a novel influenza vaccine. Sci Rep 8(1):18007

Song JM, Hossain J, Yoo DG, Lipatov AS, Davis CT, Quan FS, Chen LM, Hogan RJ, Donis RO, Compans RW, Kang SM (2010) Protective immunity against H5N1 influenza virus by a single dose vaccination with virus-like particles. Virology 405(1):165–175

Song JM, Van Rooijen N, Bozja J, Compans RW, Kang SM (2011) Vaccination inducing broad and improved cross protection against multiple subtypes of influenza a virus. Proc Natl Acad Sci U S A 108(2):757–761

Song BM, Kang HM, Lee EK, Jung SC, Kim MC, Lee YN, Kang SM, Lee YJ (2016) Supplemented vaccination with tandem repeat m2e virus-like particles enhances protection against homologous and heterologous hpai h5 viruses in chickens. Vaccine 34(5):678–686

Thrane S, Janitzek CM, Matondo S, Resende M, Gustavsson T, de Jongh WA, Clemmensen S, Roeffen W, van de Vegte-Bolmer M, van Gemert GJ, Sauerwein R et al (2016) Bacterial superglue enables easy development of efficient virus-like particle based vaccines. J Nanobiotechnology 14:30

Tretyakova I, Hidajat R, Hamilton G, Horn N, Nickols B, Prather RO, Tumpey TM, Pushko P (2016) Preparation of quadri-subtype influenza virus-like particles using bovine immunodeficiency virus gag protein. Virology 487:163–171

Urakami A, Sakurai A, Ishikawa M, Yap ML, Flores-Garcia Y, Haseda Y, Aoshi T, Zavala FP, Rossmann MG, Kuno S, Ueno R et al (2017) Development of a novel virus-like particle vaccine platform that mimics the immature form of alphavirus. Clin Vaccine Immunol 24(7): e00090–17

Valero-Pacheco N, Pérez-Toledo M, Villasís-Keever M, Núñez-Valencia A, Boscó-Gárate I, Lozano-Dubernard B, Lara-Puente H, Espitia C, Alpuche-Aranda C, Bonifaz LC, Arriaga-Pizano L et al (2016) Antibody persistence in adults two years after vaccination with an h1n1 2009 pandemic influenza virus-like particle vaccine. PLoS ONE 11(2):e0150146

Viboud C, Gostic K, Nelson MI, Price GE, Perofsky A, Sun K, Sequeira Trovão N, Cowling BJ, Epstein SL, Spiro DJ (2020) Beyond clinical trials: evolutionary and epidemiological considerations for development of a universal influenza vaccine. PLoS Pathog 16(9):e1008583

Walpita P, Johns LM, Tandon R, Moore ML (2015) Mammalian cell-derived respiratory syncytial virus-like particles protect the lower as well as the upper respiratory tract. PLoS ONE 10(7): e0130755

Wang JL, Zhang NZ, Li TT, He JJ, Elsheikha HM, Zhu XQ (2019) Advances in the development of anti-toxoplasma gondii vaccines: challenges, opportunities, and perspectives. Trends Parasitol 35(3):239–253

Ward BJ, Landry N, Trépanier S, Mercier G, Dargis M, Couture M, D'Aoust MA, Vézina LP (2014) Human antibody response to n-glycans present on plant-made influenza virus-like particle (vlp) vaccines. Vaccine 32(46):6098–6106

Weiss DJ, Lucas TCD, Nguyen M, Nandi AK, Bisanzio D, Battle KE, Cameron E, Twohig KA, Pfeffer DA, Rozier JA, Gibson HS et al (2019) Mapping the global prevalence, incidence, and mortality of plasmodium falciparum, 2000–17: a spatial and temporal modelling study. Lancet 394(10195):322–331

Wetzel D, Chan JA, Suckow M, Barbian A, Weniger M, Jenzelewski V, Reiling L, Richards JS, Anderson DA, Kouskousis B, Palmer C et al (2019) Display of malaria transmission-blocking antigens on chimeric duck hepatitis B virus-derived virus-like particles produced in *Hansenula polymorpha*. PLoS ONE 14(9):e0221394

Whitacre DC, Espinosa DA, Peters CJ, Jones JE, Tucker AE, Peterson DL, Zavala FP, Milich DR (2015) P. Falciparum and P. Vivax epitope-focused VLPS elicit sterile immunity to blood stage infections. PLoS One 10(5):e0124856

Yenkoidiok-Douti L, Williams AE, Canepa GE, Molina-Cruz A, Barillas-Mury C (2019) Engineering a virus-like particle as an antigenic platform for a pfs47-targeted malaria transmission-blocking vaccine. Sci Rep 9(1):16833

Zakeri B, Fierer JO, Celik E, Chittock EC, Schwarz-Linek U, Moy VT, Howarth M (2012) Peptide tag forming a rapid covalent bond to a protein, through engineering a bacterial adhesin. Proc Natl Acad Sci U S A 109(12):E690-697

Zambon MC (1999) Epidemiology and pathogenesis of influenza. J Antimicrob Chemother 44 Suppl B:3–9

Zeltins A (2013) Construction and characterization of virus-like particles: a review. Mol Biotechnol 53(1):92–107

Zhang X, Wei M, Pan H, Lin Z, Wang K, Weng Z, Zhu Y, Xin L, Zhang J, Li S, Xia N et al (2014) Robust manufacturing and comprehensive characterization of recombinant hepatitis e virus-like particles in hecolin(®). Vaccine 32(32):4039–4050

Zhang S, Akmar LZ, Bailey F, Rath BA, Alchikh M, Schweiger B, Lucero MG, Nillos LT, Kyaw MH, Kieffer A, Tong S et al (2020) Cost of respiratory syncytial virus-associated acute lower respiratory infection management in young children at the regional and global level: a systematic review and meta-analysis. J Infect Dis 222(Supplement_7):S680–S687

# Protein and Peptide Nanocluster Vaccines

Timothy Z. Chang and Julie A. Champion

## Contents



**Abstract** Recombinant protein- and peptide-based vaccines can deliver large amounts of specific antigens for tailored immune responses. One class of these are protein and peptide nanoclusters (PNCs), which are made entirely from the cross-linked antigen. PNCs leverage the inherent immunogenicity of nanoparticulate

T. Z. Chang
Department of Systems Biology, Blavatnik Institute at Harvard Medical School, Boston, MA 02115, USA

Wyss Institute for Biologically Inspired Engineering, Harvard University, Boston, MA 02115, USA

J. A. Champion (✉)
School of Chemical and Biomolecular Engineering, Georgia Institute of Technology, Atlanta, GA 30332, USA
e-mail: julie.champion@chbe.gatech.edu

Current Topics in Microbiology and Immunology (2021) 433: 107–130
https://doi.org/10.1007/82_2020_228

Published Online: 10 November 2020

antigens while minimizing the use of excipients normally used to create them. In this chapter, we discuss PNC fabrication methods, immunostimulatory properties of nanoclusters observed in vitro and in vivo, and protective benefits of PNC vaccines against influenza and cancer mouse models. We conclude with an outlook on future studies of PNCs and PNC design strategies, as well as their use in future vaccine formulations.

## 1 Introduction

Generating a successful immune response involves not only delivering antigen to the immune system but presenting that antigen in an immunostimulatory context. To accomplish this, the subunit vaccine nanoparticle design generally follows two strategies: (i) internal encapsulation of antigen and/or (ii) native antigen display on a particle surface. In the first strategy, the antigen is blended into the nanoparticulate polymer matrix. Interaction of the nanoparticle with the immune cells, together with the controlled release of the antigen from the nanoparticle, result in a stronger stimulation of the immune system compared to the soluble antigen (Gregory et al. 2013; Singh 2007; Liu et al. 2016). In the second strategy, the antigen is attached to the nanoparticle surface instead of being blended into the polymer matrix. This enables interaction of the particle with immune cells in a manner that leads to a superior immune response. For example, virus-like particles display multimeric epitopes in native conformations, enhancing the quality and quantity of the humoral immune response (Jegerlehner et al. 2002; Roldão et al. 2010). Protein and peptide nanocluster (PNC) vaccines capture the advantages of both these strategies. PNCs are composed almost entirely of antigen protein or peptide. These nanoclusters deliver large quantities of the antigen to antigen-presenting cells (APCs) and also display epitopes on their surface, which results in a better immune response. Unlike other vaccine nanoparticles, PNCs are made entirely of biodegradable, crosslinked antigen, minimizing the possibility of off-target immune responses. Furthermore, the peptide degradation products of protein nanoclusters are capable of contributing to an immune response, through their presentation on the major histocompatibility (MHC) proteins (Tsoras and Champion 2018). The lack of a lipid bilayer membrane in PNCs, such as that found in VLP-based vaccines, reduces manufacturing complexity and also enhances the resilience of PNCs to osmotic stresses brought on by cold chain-independent storage, a desirable property for vaccine transportation to the developing world (Chang et al. 2018).

Similar protein-based nanoparticle vaccines have been developed that use self-assembly motifs to create geometrically well-defined repeats of antigen. Kanekiyo et al. fused influenza hemagglutinin (HA) to a 24-mer self-assembling ferritin core, which resulted in 20 nm nanoparticles that could protect against an influenza challenge (Kanekiyo et al. 2013). The self-assembling protein nanoparticles (SAPNs) developed by the Burkhard group also display between 20 and 60 copies of fused protein antigens (Karch et al. 2018). One concern with these designs is the immunogenicity of the self-assembly tags. Kanekiyo et al. showed that immunization with HA-ferritin nanoparticles did not induce an immune response to endogenous host ferritin, but antibodies against the core ferritin were generated. Immune responses against the self-assembly motif in these particles could preclude strong antigen-specific responses because of original antigenic sin. This phenomenon, in which the immune system preferentially generates antibodies to epitopes previously encountered instead of new ones, needs to be addressed specifically in the case of immunizations requiring multiple boosts (Murphy et al. 2012).

Since PNC vaccines can direct nanoparticle formation without the use of an extra self-assembly tag on the antigen, PNC vaccines possess reduced chances of stimulating an off-target immune response (Willett et al. 2004). However, a robust and high-yield method for nanocluster formation must be established in the absence of engineered self-assembly. Our lab has found that desolvation is a simple, yet efficient method for generating protein and PNC. Based on an assortment of results obtained from the use of model and disease-specific proteins and peptides as antigens, PNCs have proven to be effective vaccines, inducing robust humoral and cellular responses in mice and protecting against lethal viral challenges.

## 2 Nanoclusters by Desolvation

Desolvated PNCs are formed entirely from protein or peptide by the solvent-directed assembly (Weber et al. 2000). In desolvation, an unfavorable solvent is introduced into a protein solution to increase solute-solute interactions, causing proteins or peptides in solution to coalesce into nanoparticles (Fig. 1).

Both proteins and small peptides can be desolvated into PNCs, depending on the specific antigens of interest. In some instances, immune responses to only particular domains of a protein are desired. In the case of the influenza hemagglutinin (HA) protein, the variable head region is immunodominant, while the conserved stalk region is harder to raise antibodies (Zhang et al. 2019). Antibodies against the HA stalk are cross-protective against multiple influenza strains (Kallewaard et al. 2016; Krammer and Palese 2013), and vaccine nanoparticles containing only the stalk region of HA are one strategy for eliciting those antibodies specifically (Deng et al. 2018a, b). In cancer vaccines, the antigen is usually of host origin and contains many non-immunogenic epitopes. In these cases, designing an immunogenic antigen requires enriching for the immunodominant epitopes (Li et al. 2014). Combined with the fact that T cell responses to cancer vaccines are essential, minimal peptide epitopes are an ideal antigen for cancer PNC vaccines (Tsoras and Champion 2018). In general, MHC I-restricted presentation of peptides to T cells requires least eight amino acids in length, while MHC II-restricted presentation requires peptides of at least 13 amino acids (Murphy et al. 2012). Peptides smaller than this length are not recommended as antigens.

Following antigen identification, a solvent must be selected. For large protein antigens, this is typically phosphate-buffered saline (PBS), as the goal is to keep the

**Fig. 1** Scanning electron micrographs of ovalbumin protein nanoclusters (Chang et al. 2016) Reproduced from Chang et al. (2016) with permission from the Royal Society of Chemistry

protein folded and soluble in the solvent and, ultimately, preserve folding during the nanocluster formation process. The solvent pH should avoid the isoelectric point of the protein, as this could cause premature aggregation. Solution pH is also important for antigens that undergo conformational changes in the endosome. Many viral coat proteins responsible for facilitating endosomal escape undergo drastic, pH-dependent conformational changes (Russell et al. 2018; Kirchdoerfer et al. 2018). As a result, low pH can produce conformational antigens not normally found on extracellular viruses, reducing the quality of the induced immune response. For small peptide antigens, there are more choices for solvent as peptide solubility varies widely based on the sequence and often peptide antigens have little or no secondary structure. Hexafluoroisopropanol (HFIP) is most commonly used for peptides as it is used in solid-state peptide synthesis due to its ability to solubilize most peptides, regardless of sequence. This solvent is quite dangerous to work with, however, and extreme care must be taken.

The soluble protein or peptide antigen is then desolvated into nanoclusters by the slow addition of desolvent to antigen under stirring. The thermodynamically unfavorable interaction between the desolvent and the proteins or peptides forces them to cluster together into nanoaggregates of hundreds of nanometers in size (Fig. 2a). The desolvent needs to be miscible with the solvent to allow solvent–desolvent interactions to overcome solvent-protein interactions (Fig. 2b). A similar principle underlies "salting out" of proteins in protein purification, and sodium chloride was

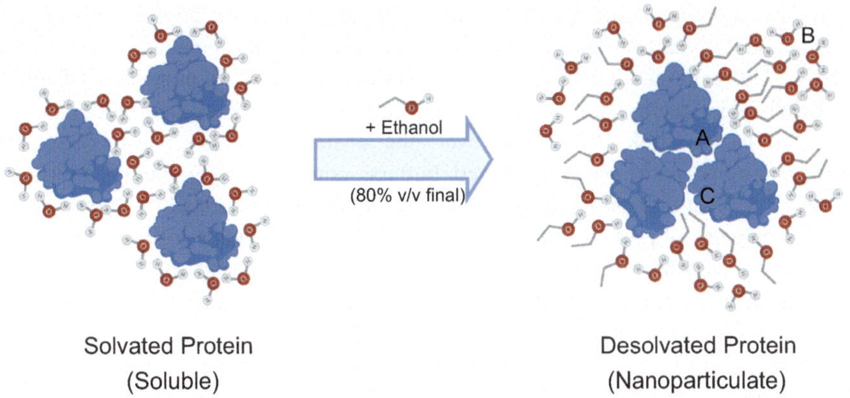


**Fig. 2** Desolvation of protein into nanoclusters involves the addition of ethanol, or another water-miscible solvent, to a protein solution. Nanocluster formation relies in part on (**a**) stabilizing protein-protein interactions induced by unfavorable protein-desolvent interactions, and (**b**) solvent sequestration or "salting out" by the desolvent. The desolvent can also stabilize hydrophobic protein domains (**c**), which can lead to the denaturation of surface-exposed epitopes. Figure created with Biorender.com

also used as an early desolvent (Kreuter 1991). For large protein antigens, ethanol is typically used as the desolvent for PBS, while diethyl ether is used with HFIP for peptide antigens. Other desolvents for proteins in PBS include acetone (Langer et al. 2003), acetonitrile (Mohammad-Beigi et al. 2016), methanol (Mohammad-Beigi et al. 2016; Doan and Ghosh 2019), and ethanol/methanol blends (Storp et al. 2012). Desolvent choice has a significant effect on protein nanocluster size (Mohammad-Beigi et al. 2016; Storp et al. 2012). While proteins generally undergo desolvation to a final concentration of 80% ethanol by volume, peptides require a much higher ratio of desolvent to solvent (Tsoras and Champion 2018), possibly owing to their higher solvent accessibility.

Following desolvation, a crosslinker is added to covalently stabilize the nascent nanoparticles and prevent aggregation or disintegration. Glutaraldehyde is a commonly used crosslinker to link primary amines, such as those found in lysines and the N termini of proteins and peptides. While glutaraldehyde has been used in the past to crosslink protein nanoparticles (Estrada et al. 2014; Wang et al. 2014), our work also uses the crosslinker 3,3′-dithiobis[sulfosuccinimidylpropionate], or DTSSP. DTSSP is also an amine-reactive crosslinker but contains a central disulfide bond that has the potential to be reduced once the particle is inside the cell. DTSSP crosslinking of PNCs in the presence of desolvent or pure solvent modestly affects PNC size (Chang et al. 2016), while glutaraldehyde concentration and crosslinking time strongly influence PNC size (Baseer et al. 2019). Though bifunctional crosslinkers are sufficient for stabilizing protein nanoclusters, trifunctional crosslinkers are needed for PNC. This is likely due to the much smaller size and number of reactive groups in peptides compared to larger proteins. Similar to bifunctional crosslinkers, both non-reducible and reducible trifunctional crosslinkers are available that react with amine or thiols in the peptides. Genipin, a compound derived from the *Genipa americana* fruit, has also been used to crosslink PNCs (Dong et al. 2019). To couple carboxyl groups to amines, EDC/NHS (1-Ethyl-3-(3-dimethylaminopropyl) carbodiimide/N-hydroxysuccinimide) can also be used to crosslink PNCs. However, the unstable reaction intermediate and incompatibility with phosphate-based buffers, combined with the instability of some proteins at the optimal pH 6 for this reaction make this a less attractive option.

The delivery of properly folded protein antigen is especially desirable in protein nanocluster vaccine design. Desolvation of proteins can lead to denaturation of protein on the surface of the nanocluster (Fig. 2c), which could diminish antigen recognition. In addition to less efficient antigen presentation, exposure of denatured epitopes can alter the protein corona or layer of host proteins that coat the surface of any nanoparticle administered in vivo (Lindman et al. 2007; Fleischer and Payne 2014; Ezzat et al. 2019). To avoid a sub-optimal immune response, we have found that coating protein nanoclusters with an additional layer of protein, either antigen (Wang et al. 2017) or adjuvant (Chang et al. 2017), enhances the immune response. While coating may not be necessary for vaccine nanoclusters made of small antigens or peptides (Tsoras and Champion 2018; Wang et al. 2014), it is useful for presenting conformational antigens to the humoral immune system (Bergtold et al. 2005). Surface antigen display on vaccines is emerging as a trend in other vaccine designs (Gregory et al. 2013; Xiang et al. 2013), as opposed to viewing particulates as mere antigen depots. This shift has been driven by two insights: (1) an immunological understanding that surface receptor engagement on antigen-presenting cells (APC) is essential for optimal interfacing with the innate and adaptive immune systems (Zhao et al. 2014), and (2) the discovery that APC engagement with nanoparticles themselves triggers inflammatory responses (Li et al. 2008; Hornung et al. 2008; Sharp et al. 2009).

Protein nanoclusters have been made from abundant proteins such as human and bovine serum albumin and gelatin since as early as the 1970s (Kreuter 1991; Marty et al. 1978). Even today, desolvated protein nanoclusters are made from albumin (Weber et al. 2000; Langer et al. 2003; Amighi et al. 2020) and gelatin (Subara et al. 2017; Jahanshahi 2008), have the most extensively characterized synthesis parameters of any protein nanoclusters. However, our lab has demonstrated that protein nanoclusters can also deliver active therapeutic enzymes to cells (Estrada et al. 2014; Herrera Estrada et al. 2017). Because of albumin's high affinity for hydrophobic small molecules (Al-Husseini et al. 2019), and its low antigenicity, albumin nanoclusters are an attractive drug delivery vehicle for enhancing small molecule half-life and biodistribution (Keuth et al. 2020; Luebbert et al. 2017). These nanoclusters are readily taken up by macrophages (Langer et al. 2003) and this can be used to target small molecule drugs to these cells in particular (Markovsky et al. 2007).

The same properties that make desolvated protein nanoclusters attractive to macrophages make them ideal vehicles for vaccine delivery. Nanoparticulate antigen is more immunogenic than soluble antigen (Gregory et al. 2013; Singh 2007; Wang et al. 2014), and targeting it to phagocytic APCs enhances immunogenicity. In our studies of protein nanocluster vaccines, we demonstrate robust in vitro dendritic cell responses to PNCs, enhanced retention in draining lymphatic organs, and successful immunization with a variety of protein and peptide antigens.

# 3 Functional Benefits of Protein Nanocluster Vaccines

The particulate formulation of protein nanoclusters provides distinct advantages over soluble antigens, as summarized in Fig. 3. Some of these advantages are general to all nanoparticle vaccines and some are specific to crosslinked protein nanoclusters.

## 3.1 Delivery

All nanoparticles, including nanoclusters, benefit from improved delivery due to their relatively large size compared to soluble antigen. Upon intradermal (i.d.) or intramuscular (i.m.) administration, nanoclusters are retained at the injection site significantly longer than soluble antigen (Tsoras and Champion 2018; Deng et al.

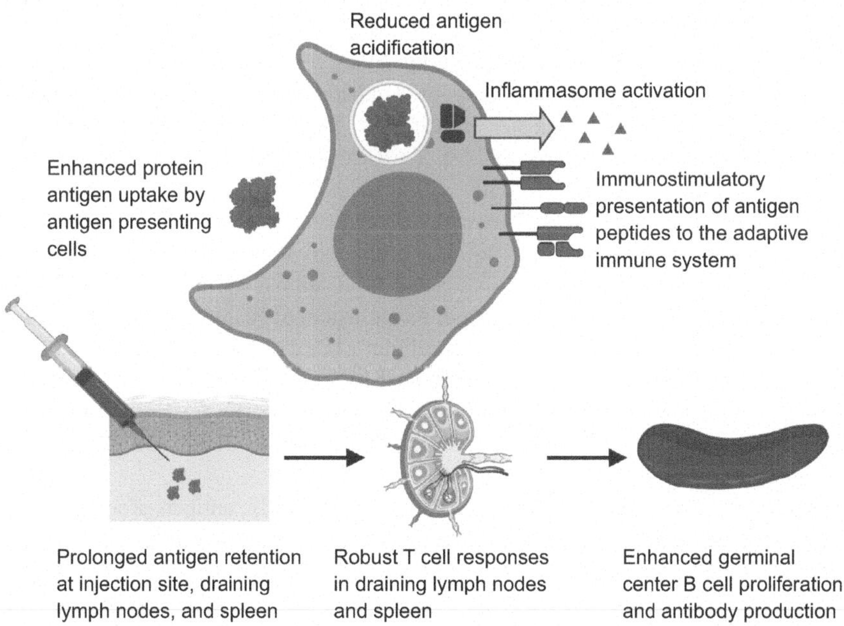


**Fig. 3** Summary of benefits of protein and PNC vaccines observed at cellular and physiological levels. Figure created with Biorender.com

2018b). The presence of nanoclusters for longer than 2 and 5 days for i.d. and i.m. injection, respectively, indicates that immune cells in the tissue are persistently exposed to antigen and increases the likelihood of antigen uptake. Nanoclusters also exhibit different trafficking and biodistribution compared to soluble antigen. When PNC was administered i.d., the antigen was detected in draining lymph nodes for 4–24 h after injection, which can increase the likelihood of presentation to T cells and activation (Tsoras et al. 2020). Nanocluster peptide antigen was not detected in the spleen, and soluble peptide antigen was detected in the spleen only transiently. The route of administration is important for antigen trafficking, as protein nanoclusters administered i.m. exhibited antigen trafficking to both the draining lymph nodes and spleen that was detectable more than a week after injection (Deng et al. 2018a, b). Antigen accumulation in the spleen was significantly higher for protein nanoclusters than soluble protein antigen, though levels in the draining lymph nodes were similar. The diameter of most antigen nanoclusters is in the range of 200 nm, which is the upper limit seen for nanoparticles that traffic directly in the lymphatic system (Reddy et al. 2007; Manolova et al. 2008). Likely, most nanoclusters are taken up by APCs at the injection site and are actively trafficked through the lymphoid system whereas soluble antigens would diffuse directly.

## 3.2 *Antigen Presentation*

Given the importance of antigen uptake by APCs for both trafficking and presentation, enhanced uptake of nanoclusters by dendritic cells (DCs) also contributes to their overall function. Nanoclusters made from large protein antigens show significant increases in uptake by DCs in vitro, regardless of the size of the nanoclusters (Chang et al. 2016). This is consistent with other types of vaccine nanoparticles (Snapper 2018). However, nanoclusters made from small peptide antigens (minimal epitopes) did not show any advantage in uptake relative to soluble small peptides, with both exhibiting high uptake by DCs in vitro (Tsoras and Champion 2018). Once internalized, nanoclusters appeared to traffic differently than soluble antigen in cells, as the intracellular pH experienced by soluble antigen was significantly lower than that experienced by nanoclusters. Reduced acidification can decrease antigen degradation, which may lead to improved cross-presentation, and could indicate endosomal escape (Tran and Shen 2009; Accapezzato et al. 2005).

While uptake by APCs is a critical first step, ultimately, antigen processing and presentation are needed to activate antigen-specific T cells (Murphy et al. 2012). For MHC I peptide antigens for which presentation can be measured, it was observed that DCs exposed to nanoclusters demonstrated greater levels of presentation than those exposed to soluble antigen both in vitro and in draining lymph nodes following i.d. administration in vivo (Tsoras and Champion 2018; Tsoras et al. 2020). Concurrently, nanoclusters induced increased DC display of maturation factors CD80 and CD86 compared to soluble antigen when incubated with DCs in vitro (Tsoras and Champion 2018; Chang et al. 2016). This data demonstrates that even when there is little or no enhancement in antigen uptake via nanoclusters, the subsequent steps of presentation and maturation are more productive for nanoclusters than soluble antigen.

Similar to maturation markers, in vitro DC secretion of inflammatory cytokine IL-1β was increased for nanoclusters containing either ovalbumin (OVA) or influenza antigens matrix protein 2 ectodomain (M2e) or M2e and HA (Chang et al. 2016; Deng et al. 2017). IL-1β is a product of inflammasome activation and induces rapid cleavage of pro-inflammatory cytokines into their active form, triggering a local, innate immune response (Murphy et al. 2012). Not all vaccine nanoparticles are capable of triggering inflammasome activation (Neumann et al. 2014; Gross et al. 2011). Interestingly, in vitro DC secretion of inflammatory cytokine TNF-α was lower for OVA nanoclusters than soluble OVA and higher for HA or M2e nanoclusters compared to soluble antigen (Chang et al. 2016; Deng et al. 2017, 2018a, b), though the role of TNF-α in vaccine responses is not clear (Murphy et al. 2012).

In vivo maturation of DCs in response to SIINFEKL nanoclusters mirrored that seen in vitro. SIINFEKL is an MHC I minimal peptide epitope from OVA protein. Upon i.d. vaccination, higher levels of double-positive, antigen-presenting, and CD86 positive, DCs were observed in the draining lymph nodes for SIINFEKL nanoclusters than for soluble antigen (Tsoras et al. 2020). Unexpectedly, this was only true for nanoclusters that were crosslinked with a non-reducible crosslinker that formed amide bonds between antigens, which is incidentally the same bond cleaved by proteases (Kisselev et al. 2000; Hedstrom 2002). Nanoclusters crosslinked with a reducible crosslinker, or a crosslinker that was non-reducible and also not cleavable by proteases did not induce significant increases in DC antigen presentation and CD86 expression. We speculate that proteolytic cleavage of the nanoclusters might have enhanced presentation and maturation. Large proteins are required to be proteolytically processed by APCs, whether or not they are in nanoclusters, to extract epitopes for MHC presentation. Antigens in nanoclusters must be similarly extracted by the cellular machinery for MHC presentation. By providing crosslinks between antigens that can be readily degraded by the natural protein-degradation machinery of the cells, nanoclusters may promote antigen extraction and MHC presentation. Importantly, in vivo DC maturation was antigen-specific as the increase in CD86 positive DCs was only seen for those DCs presenting antigen, and, the increases in the fraction of DCs presenting antigen were only seen for CD86 positive DCs (Tsoras et al. 2020). This was true of all nanoclusters, regardless of the crosslinker type. Administration of soluble antigen with poly (I:C), a synthetic double-stranded RNA adjuvant, upregulated CD86 in DCs that did not present antigen, demonstrating non-specific activation that is not necessarily productive to the adaptive immune response and may be responsible for side effects.

### 3.3 Cellular Responses

Following successful antigen presentation, CD4 and CD8 T cells are activated. This activation depends on the type of antigen and MHC used for presentation (Murphy et al. 2012). Protein and PNC have demonstrated antigen-specific T cell activation with several different antigens and in different tissues, depending on the route of administration. M2e nanoclusters induced a strong systemic and mucosal cellular response upon intranasal (i.n.) vaccination, as cells collected from both the spleen and lungs secreted IFN-$\gamma$ and IL-4 in response to M2e restimulation (Wang et al. 2014). Influenza nucleoprotein (NP) nanoclusters and nanoclusters made from peptides derived from NP were coated with M2e and administered i.m. The coated nanoclusters, but not soluble mixtures of the proteins, elicited antigen-specific responses to both M2e and NP. Upon restimulation, splenocytes from mice administered nanoclusters secreted IFN-$\gamma$, IL-4, and IL-2 at much higher levels than those from animals administered soluble protein (Deng et al. 2018b). Similarly, HA-coated M2e nanoclusters also induced specific cellular responses, as demonstrated by high IFN-$\gamma$ secretion following restimulation.

The responses described above were for nanoclusters without added adjuvant, where the nanocluster itself served as the delivery vehicle and "self-adjuvant". We have also investigated the effect of adjuvants co-delivered with nanoclusters, as coatings, incorporated within the nanoclusters, or as simple mixtures. Incorporation of CpG oligonucleotide, a toll-like receptor 9 agonists, within M2e nanoclusters did not have any effect on cellular or antibody responses (Wang et al. 2014). M2e and the H2 domain of HA have also been incorporated as fusion proteins within the variable domain of bacterial adjuvant flagellin (flic) (Deng et al. 2017). Flic is a toll-like receptor 5 agonist and has been demonstrated to boost anti-influenza immune responses (Kim et al. 2015; Oh et al. 2014). Flic-M2e nanoclusters and flic-H2HA coated flic-M2e nanoclusters were formed by crosslinking without desolvation and administered to mice i.n.. All nanoclusters induced high levels of M2e-specific cellular responses, measured by IL-2 secretion upon restimulation ex vivo. Though not typically considered as adjuvants, immunoglobulins from the host can serve this purpose. Anti-OVA IgM coated on the surface of OVA nanoclusters induced the formation of more central memory T cells relative to OVA-coated OVA nanoclusters (Chang et al. 2017).

Using PNCs made from SIINFEKL, a CD8 T cell antigen, we have more deeply investigated the T cell response following i.d. administration. Relative to soluble SIINFEKL, PNCs showed increased expression of CD69, an early marker of T cell activation, on CD8 T cells from draining lymph nodes, corresponding to the previously described evidence of DC presentation in those nodes (Tsoras et al. 2020). Adjuvanting soluble SIINFEKL with poly(I:C) resulted in increased CD69 expression in CD8 and CD4 T cells in both the draining lymph nodes and spleen, indicating non-specific activation. Only nanoclusters crosslinked with amide bonds, those which showed enhanced DC presentation and maturation, induced high secretion of IFN-$\gamma$ upon restimulation of lymph node CD8 T cells, equivalent to poly(I:C) adjuvanted peptide. This connection between nanocluster processing, presentation, and T cell activation is encouraging and will motivate future work on the role of antigen crosslinking and cleavage in nanoclusters. Altogether, across a wide combination of antigens, nanoclusters reliably induce a strong cellular immune response that is, importantly, antigen-specific.

## 3.4 Humoral Responses

In addition to cellular responses, PNCs are capable of stimulating robust humoral responses. Nanoclusters of a variety of designs presenting B cell epitopes from M2e and HA induced significant antibody production. The repetitive, multi-valent presentation of antigens on the surface of nanoclusters is likely beneficial for recognition by B cell receptors (Wang et al. 2012; Schellekens and Jiskoot 2013; Kim et al. 2006). M2e nanoclusters induced a strong humoral response upon i.n. vaccination, including serum antibody subtypes IgG1, IgG2a, and IgG2b indicative of balanced Th1 and Th2 responses (Wang et al. 2014). Further, mucosal antibody response was seen as lung and nasal washes revealed the production of both IgG and IgA. When M2e was coated on the surface of NP nanoclusters and administered i.m., a similar strong serum M2e-specific antibody response was seen (Deng et al. 2018a, b). No NP-specific antibodies were detected. However, when nanoclusters made from M2e were coated with HA, M2e-specific antibodies were still produced at high levels (Deng et al. 2018b). There was a small, but statistically significant, decrease in M2e antibody titers for HA-coated nanoclusters compared to uncoated nanoclusters. Whether immunological differences between M2e and NP antigens are responsible, or if the HA coating did not completely obscure M2e on the surface of the particle, is not clear. For HA specifically, a crosslinked coating of HA was required on the surface of nanoclusters to induce antibody production. This is likely due to the sensitive nature of the structure of trimeric HA, which was disrupted in the desolvation process. HA-coated HA nanoclusters exhibited high titers of IgG, including IgG1, IgG2a, and IgG2b, though the response was biased toward Th2 (Wang et al. 2017). The antibodies were shown to be neutralizing and inhibit hemagglutination. Serum antibody titers were dependent on the dose and route of administration, with 2 i.m. doses inducing far higher levels than 1 i.m., 1 i.n., or 2 i. n. doses. Despite the large difference in antibody response, 2 doses given either i.m. or i.n. both protected fully from weight loss and death upon H7N9 viral challenge. Similarly, when M2e nanoclusters were coated with trimeric HA, HA-specific antibody responses were strong (Deng et al. 2018a, b).

The humoral responses described above were all the result of nanocluster delivery of antigen with no external adjuvant provided. In the case of flic-M2e fusion protein nanoclusters and flic-H2HA fusion coated flic-M2e nanoclusters, strong antibody responses were induced that were specific to both M2e and HA (Deng et al. 2017). Interestingly, incorporation of the flic-HA coating improved Me2-specific serum IgG titers. In a different configuration, we found that flagellin-coated OVA nanoclusters and soluble flagellin mixed with OVA nanoclusters induced similar levels of antibodies upon i.m. immunization. However, flagellin-coated nanoclusters induced significant affinity maturation of antibodies, while flagellin-admixed nanoclusters did not (Chang et al. 2017). Anti-OVA IgM coated on the surface of OVA nanoclusters increased IgG1 and IgG2a serum titers relative to OVA-coated OVA nanoclusters (Chang et al. 2017). Interestingly, the IgM coating prevented affinity maturation despite its induction of class switching. Affinity maturation was also observed for OVA-coated nanoclusters, which is consistent with our data showing splenic germinal center B cell proliferation upon immunization with HA-coated HA nanoclusters (Chang 2017). The reason for adjuvant-dependent differences in affinity maturation is not clear. Affinity maturation has not been studied for non-model antigens, such as influenza proteins.

### *3.5 Storage and Stability*

There are many challenges for global vaccination campaigns. One significant challenge is the need for cold chain storage of vaccines, especially in developing countries with unreliable electricity and transportation infrastructure as well as hot climates. We assessed the ability of HA-coated HA nanoclusters to maintain their potency over storage time, as measured by hemagglutination activity in vitro, and antibody titers and hemagglutination inhibition titers following i.m. vaccination. Nanoclusters aged in solution (PBS) for 3 months at 25 °C retained their full immunogenicity, relative to fresh nanoclusters (Chang et al. 2018). M2e coated NP nanoclusters retained full activity after 2 months storage at 4 °C and had a small, but statistically significant, loss after 2 months storage at 25 °C, as measured by endpoint antibody titers (Deng et al. 2018b). Stability may depend on the exact antigen used and is also likely to be affected by the amount of crosslinking, protein concentrations, and any formulation stabilizers. These factors will be important to investigate thoroughly in future work.

## 4 Application of Protein Nanoclusters as Protective Vaccines

### *4.1 Broadly Protecting Influenza Vaccines*

Seasonal influenza vaccines provide limited protection, which is better is some years and worse in others, based, in part, on how well the strains included in the vaccine match the circulating strains in the community (CDC CDC Seasonal Flu Vaccine Effectiveness Studies 2020). Additionally, zoonotic influenza strains, such as the recent avian strains, do infect humans occasionally and are serious health

**Table 1** M2e sequences from different human influenza A strains and the consensus sequence (Wang et al. 2014)

| Viral strains | Subtype | M2e amino acid sequence |
|---|---|---|
| M2e in nanoclusters | N/A[a] | MSLLTEVETP IRNEWGCRCN D |
| A/Philippines/2/82 | H3N2 | MSLLTEVETP IRNEWGCRCN D |
| A/Puerto Rico/8/34 | H1N1 | MSLLTEVETP IRNEWGCRCN **G** |
| A/California/04/09 | H1N1 | MSLLTEVETP**T**R**S**EW**E**CRC**S**D |
| A/Vietnam/1230/04 | H5N1 | MSLLTEVETP**T**RNEW**E**CRC**S**D |

Reproduced from Wang et al. (2014)

[a]M2e consensus of human influenza A viruses

threats (Lai et al. 2016; Gao et al. 2013; Wu et al. 2016). A universal influenza vaccine is needed that can protect populations against seasonal and pandemic influenza strains, including future strains that have not yet evolved. One challenge of universal influenza vaccines is the high mutation rate of influenza, specifically in the immunodominant antigens, hemagglutinin, specifically the head domain, and neuraminidase (Altman et al. 2018). The highly conserved ectodomain of matrix protein 2 is a promising antigen for a universal influenza vaccine (Lamb et al. 1985), however, its immunogenicity is low. Though M2e specific antibodies can reduce viral load, they are rarely detected after natural infection or seasonal vaccination (Hughey et al. 1995; Treanor et al. 1990; Black et al. 1993; Feng et al. 2006). The first application of PNCs sought to improve the immunogenicity of M2e and evaluate the breadth of protection. We selected the M2e consensus sequence from human influenza A viruses as the antigen, as shown in Table 1.

To recapitulate the native tetrameric state of M2e, the GCN4 tetramerization motif, modified from a leucine zipper region of a yeast transcription factor (De Filette et al. 2008), was fused to M2e and expressed in insect cells (Wang et al. 2014). Nanoclusters were fabricated by desolvating the tetrameric M2e complex and crosslinking with non-reducible glutaraldehyde. Mice were vaccinated i.n. with 10 μg of M2e each in a prime and two boost doses. The M2e specific humoral and cellular responses generated by vaccination with nanoclusters protected mice from lethal challenge. Antibodies generated recognized not only the M2e consensus peptide but mutated M2e sequences from different viral strains. While no mice immunized with soluble tetrameric M2e survived, all mice vaccinated with tetrameric M2e nanoclusters did survive. This protective effect was broad, as mice survived both challenged by an influenza strain with the same M2e consensus sequence (A/Philippines/82 H3N2: Phi/82 H3N2) as the nanoclusters, and challenged by a divergent strain with four mutated residues out of 21 (A/California/04/09 H1N1: CA/09 H1N1). While survival was promising, the mice did lose weight, up to 10 and 20% for similar and divergent strains, respectively, indicating the need to further enhance the vaccine to improve the quality and breadth of protection. Passive transfer of immune sera from nanocluster-immunized mice to naïve mice protected most animals, but not all, demonstrating the importance of the cellular response in protection against influenza. There is evidence that M2e antibodies are not neutralizing but rather mediate effector cell functions (Jegerlehner et al. 2004; El Bakkouri et al. 2011; Mozdzanowska et al. 2005; Tompkins et al. 2007).

In an effort to improve the efficacy and breadth of protection, the antigen content of nanoclusters was expanded. Instead of a single M2e consensus sequence from human influenza A, a chimeric antigen was created that linked M2e consensus sequences from human, swine, avian, and fowl influenza A strain into a single chain fused to GCN4 to induce tetramerization (Fig. 4a). The goal of this modification was to improve protection against zoonotic strains. The stalk domain of HA was also used, as it is also relatively conserved, compared to the HA head domain (Ekiert et al. 2011; Mallajosyula et al. 2014) and provides additional immune epitopes. Engineered trimeric HA stalk was created from both H1 and H3 virus (Fig. 4b). Desolvation was used to create chimeric M2e nanoclusters, which were coated with H1, H3, or a mixture of H1 and H3 trimeric stalk domains and crosslinked (Deng et al. 2018a) (Fig. 4c). Upon i.m. vaccination, mice maintained early antibody titers until at least 4 months post-vaccination. The immune responses extended beyond the nanocluster-specific humoral and cellular responses seen for the antigens in the nanoclusters. Vaccination induced antibodies capable of binding not only the antigens present in the nanoclusters, but also H2 and H5 subtypes if H1 was present in the coating, and H7 and H10 subtypes if H3 was present in the coating. Broad protection was ultimately demonstrated by viral challenge with a panel of six influenza viruses: A/Puerto Rico/8/34 H1N1 (PR8 H1N1) and A/Aichi/2/1968 H3N2 (Aic H3N2), from which the H1 and H3 stalks were derived, A/California/7/2009 H1N1pdm (p09 H1N1), reassorted A/Vietnam/1203/2004 (Vtn H5N1) virus, A/Philippines/2/1982 (Phi H3N2), and reassorted A/Shanghai/2/2013 (SH H7N9) virus. The respective reassorted viruses contained HA and NA from the parent strain (Vtn or SH) but the internal genes and M2e from PR8 H1N1. In all cases, nanoclusters containing both the chimeric M2e core and the coating of either both H1 and H3 stalk or coating of the HA stalk variant most closely related to the challenge virus (H1 for H5N1 virus and H3 for H7N9 virus) fully protected mice from lethal challenge. Vaccination with only the M2e nanocluster core protected most mice, but not all. These results demonstrate high potential for a broadly protecting influenza vaccine and the benefit of combining multiple conserved antigenic domains.

To further explore the value of other conserved influenza antigens, we fabricated nanoclusters from NP and known NP peptide antigens (Deng et al. 2018b). NP is an internal influenza protein that is highly conserved and has been shown to induce cross-protection against a challenge from diverse influenza A viruses (Zheng et al. 2014). Nanoclusters were desolvated both from full-length NP and from an engineered peptide containing three linear repeats each of the $NP_{147}$ and $NP_{55}$ short peptide epitopes. Following desolvation, nanoclusters were coated with the chimeric M2e tetramer used previously (Fig. 4a) and crosslinked. Mice vaccinated i.m. with both types of M2e coated NP nanoclusters were protected from lethal challenge with the reassortant H5N1 virus. Animals did lose $\sim 10\%$ body weight and the coated nanoclusters containing NP peptide recovered faster than those

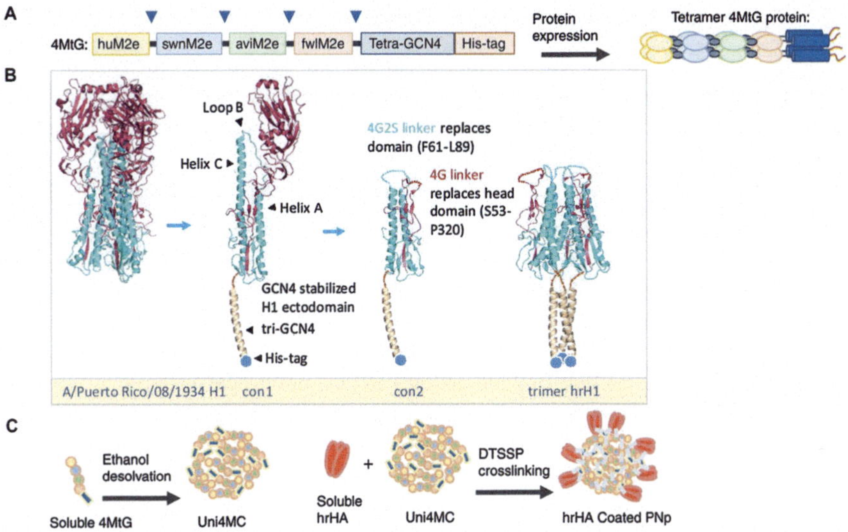


**Fig. 4** Antigen design of (**a**) chimeric M2e tetramers and (**b**) stabilized HA1 stalk antigen. Coated PNCs containing both antigens were made by the method described in Sect. 2 (**c**) (Deng et al. 2018a). Figure adapted from Deng et al. (2018a) under the Creative Commons license

containing NP protein. We also investigated whether nanoclusters could benefit from codelivery with inactivated (PR8 H1N1) influenza vaccine, as previous work had shown that codelivery of soluble flic-M2e fusion proteins and inactivated vaccines broadened the protection (Zhu et al. 2017). Using both i.m. and microneedle delivery routes, mice co-vaccinated with M2e coated NP PNC and inactivated PR8 were protected against both PR8 challenge and Phi H3N2 challenge. Mice vaccinated only with inactivated PR8 were only protected from PR8 challenge. Again, the value of antigen combination in nanoclusters is evident and there is potential for further combination with existing vaccines to boost immunogenicity.

A universal influenza vaccine would ideally protect against all potential influenza strains. However, strain-specific vaccines could be useful for enhancing responses against specific antigenic sequences in the case of an emerging outbreak. Additionally, in the case of a recombinant influenza virus engineered to evade a broadly protective vaccine, an antigen-matched formulation may be the only means of generating protective immunity. We have demonstrated that PNCs made with hemagglutinin H1 and H3 stalk domains inserted into the flic protein can trigger subtype-specific protective immunity (Deng et al. 2017). Overall, PNCs made from a variety of influenza A antigens can elicit strong, protective immune responses against homo- and heterosubtypic challenges.

### 4.2 *Progress Towards Cancer Vaccines*

Protein nanoclusters have also been recently explored as therapeutic cancer vaccines. In particular, the E.G7-OVA mouse T cell lymphoma cell line generates tumors when injected subcutaneously and can be used as a measure of OVA nanocluster efficacy. OVA nanocluster-immunized mice had drastically lower tumor volumes and growth rates compared to mice immunized with soluble OVA, which had almost no effect on tumor volume (Dong et al. 2019). CpG-coated nanoclusters also slightly, but significantly, reduced tumor volumes as compared to unadjuvanted OVA nanoclusters. Another study demonstrated the protective effects of OVA nanoclusters crosslinked with indocyanine green. This crosslinker has the bonus of generating reactive oxygen species (ROS) upon near-infrared irradiation, which was shown to be essential for endosomal escape and therapeutic effect (Cao et al. 2018).

Given the high specificity needed for cancer vaccines, many antigens are minimal peptide epitopes. We have fabricated PNCs from oncofetal antigen (OFA) peptide epitopes (Tsoras and Champion 2018). OFA is a highly conserved protein antigen expressed in many types of cancer cells including breast, head/neck, and hematologic malignancies, but it is not detectably expressed in healthy cells (Siegel et al. 2003; Barsoum and Schwarzenberger 2014). The lack of reactive groups for sufficient crosslinking required the addition of a terminal cysteine to enable stabilization of the nanoclusters using a tri thiol crosslinker. Evaluation of human cancer antigens in humanized mice or clinical trials is needed to determine efficacy, likely in combination with checkpoint inhibitors, given the highly immunosuppressive environment of many tumors (Sharma et al. 2017).

## 5 Conclusion and Future Outlook

PNCs are designed to present specific antigens and epitopes in the context of enhanced immunogenicity conferred by particulate matter. Our work has demonstrated successful immunization with PNCs, yet opportunities still remain for enhancing their immunogenicity. The immunomodulatory effects of nanoparticles have been shown to be a function of many factors, including nanoparticle size (Xiang et al. 2006; Stano et al. 2012; Mottram et al. 2007), shape (Kumar et al. 2015; Vaine et al. 2013), charge (Neumann et al. 2014), surface chemistry (Huang et al. 2013), and administration route (Zolnik et al. 2010). HSA protein nanocluster density has recently been shown to affect delivery to cells of the immune system in vitro and in vivo (Roh et al. 2019). Future vaccine nanocluster work should examine whether nanocluster density is an intrinsic property of the antigen, or if synthesis conditions can tune nanocluster density and the resulting immune response. Non-desolvated protein nanoclusters may better preserve the native structure of certain antigens and are also being explored for cancer (Dong et al. 2019; Cao et al. 2018) and influenza (Deng et al. 2017) vaccines.

We have also demonstrated the importance of the outer coat layer on protein nanoclusters for effective immune responses. The outer layer of protein nanoclusters is especially important for cell- and tissue-specific targeting. Recent studies of non-vaccine HSA nanocluster drug carriers have used small molecules (Akbarian et al. 2020), nucleic acid aptamers, and antibodies (Keuth et al. 2020) to target nanocluster delivery. The protein corona on the respiratory syncytial virus (RSV) and herpes simplex virus type 1 (HSV-I), has been shown to affect immune cell activation and viral pathogenesis (Ezzat et al. 2019). The corona of protein nanoparticle vaccines, including PNCs, has not been studied yet, though is very likely to impact biodistribution, cellular uptake, and/or immunogenicity.

The use of antibodies and other proteins of the immune system for targeting is especially relevant for vaccine nanoclusters, given that the Fc receptor is a key mediator of antigen presentation to B cells (Bergtold et al. 2005). We have tested flagellin and immunoglobulin as adjuvants to nanoclusters (Chang et al. 2017), yet other protein types should be explored. In particular, the complement protein C3d can bridge the innate and adaptive immune response by flagging particulate antigens as targets for humoral immune responses (Del Nagro et al. 2005).

In addition to humoral immune responses, viral vaccines need to induce robust cell-mediated immunity, which is a much more difficult challenge. Protein nanoclusters deliver large quantities of antigen per particle to antigen-presenting cells, increasing the chances of successful T cell epitope antigen presentation. In mice, PNCs show enhanced antigen retention in lymph nodes and the spleen (Deng et al. 2018b; Tsoras et al. 2020), important organs for antigen presentation. However, the presentation of antigen alone is not sufficient for stimulating cell-mediated immunity. T cell differentiation and polarization are mediated by cytokine transfer from the APC to the cognate T cell (Murphy et al. 2012; Munoz-Wolf and Lavelle 2018). Exogenous cytokine administration can be dangerous if not appropriately targeted (Tisoncik et al. 2012), and protein nanoclusters could provide a means of such targeting. Pathogenic pattern recognition receptor activation also leads to specific inflammatory cytokine production, and engagement of these receptors by adjuvants has been shown to be critical to successful T cell responses (Kumar et al. 2019). Whether engaging cytokine receptors directly or inducing pro-inflammatory cytokine production pathways in APCs, adjuvant-coated protein nanoclusters should be explored as a means of inducing specific cell-mediated immune phenotypes.

Though a significant benefit of PNCs is the lack of external adjuvants, PNC vaccines of the future may contain both molecular adjuvants and particulate antigens, especially as we learn more about how different host- and pathogen-derived components trigger specific aspects of the immune system. Future research should address how different adjuvant combinations, as well as antigen nanoparticle properties, can optimize immune responses to combat viral, bacterial, and/or oncological threats.

## References

Accapezzato D, Visco V, Francavilla V, Molette C, Donato T, Paroli M, Mondelli MU, Doria M, Torrisi MR, Barnaba V (2005) Chloroquine enhances human CD8(+) T cell responses against soluble antigens in vivo. J Exp Med 202(6):817–828. https://doi.org/10.1084/jem.20051106

Akbarian A, Ebtekar M, Pakravan N, Hassan ZM (2020) Folate receptor alpha targeted delivery of artemether to breast cancer cells with folate-decorated human serum albumin nanoparticles. Int J Biol Macromol 152:90–101. https://doi.org/10.1016/j.ijbiomac.2020.02.106

Al-Husseini JK, Stanton NJ, Selassie CRD, Johal MS (2019) The binding of drug molecules to serum albumin: the effect of drug hydrophobicity on binding strength and protein desolvation. Langmuir 35(52):17054–17060. https://doi.org/10.1021/acs.langmuir.9b02318

Altman MO, Angeletti D, Yewdell JW (2018) Antibody immunodominance: the key to understanding influenza virus antigenic drift. Viral Immunol 31(2):142–149. https://doi.org/10.1089/vim.2017.0129

Amighi F, Emam-Djomeh Z, Labbafi-Mazraeh-Shahi M (2020) Effect of different cross-linking agents on the preparation of bovine serum albumin nanoparticles. J Iran Chem Soc 17(5):1223–1235. https://doi.org/10.1007/s13738-019-01850-9

Barsoum AL, Schwarzenberger PO (2014) Oncofetal antigen/immature laminin receptor protein in pregnancy and cancer. Cell Mol Biol Lett 19(3):393–406. https://doi.org/10.2478/s11658-014-0203-7

Baseer A, Koenneke A, Zapp J, Khan SA, Schneider M (2019) Design and characterization of surface-crosslinked gelatin nanoparticles for the delivery of hydrophilic macromolecular drugs. Macromol Chem Phys 220(18):1900260. https://doi.org/10.1002/macp.201900260

Bergtold A, Desai DD, Gavhane A, Clynes R (2005) Cell surface recycling of internalized antigen permits dendritic cell priming of B cells. Immunity 23(5):503–514. https://doi.org/10.1016/j.immuni.2005.09.013

Black RA, Rota PA, Gorodkova N, Klenk HD, Kendal AP (1993) Antibody response to the M2 protein of influenza A virus expressed in insect cells. J Gen Virol 74(Pt 1):143–146. https://doi.org/10.1099/0022-1317-74-1-143

Cao FQ, Yan MM, Liu YJ, Liu LX, Lu L, Wang H, Zhang C, Sun HF, Kong DL, Ma GL (2018) Photosensitizer-induced self-assembly of antigens as nanovaccines for cancer immunotherapy. Biomater Sci 6(3):473–477. https://doi.org/10.1039/C7BM01082F

CDC CDC Seasonal Flu Vaccine Effectiveness Studies. https://www.cdc.gov/flu/vaccines-work/effectiveness-studies.htm (accessed 10 July 2020)

Chang TZ (2017) Protein nanoparticle vaccines. Georgia Tech

Chang TZ, Stadmiller SS, Staskevicius E, Champion JA (2016) Effects of ovalbumin protein nanoparticle vaccine size and coating on dendritic cell processing. Biomater Sci

Chang TZ, Diambou I, Kim JR, Wang B, Champion JA (2017) Host- and pathogen-derived adjuvant coatings on protein nanoparticle vaccines. Bioeng Transl Med 2(1):120–130. https://doi.org/10.1002/btm2.10052

Chang TZ, Deng L, Wang B-Z, Champion JA (2018) H7 Hemagglutinin nanoparticles retain immunogenicity after > 3 months of 25 °C storage. PLoS ONE 13(8):e0202300–e0202300. https://doi.org/10.1371/journal.pone.0202300

De Filette M, Martens W, Roose K, Deroo T, Vervalle F, Bentahir M, Vandekerckhove J, Fiers W, Saelens X (2008) An influenza A vaccine based on tetrameric ectodomain of matrix protein 2. J Biol Chem 283(17):11382–11387. https://doi.org/10.1074/jbc.M800650200

Del Nagro CJ, Kolla RV, Rickert RC (2005) A critical role for complement C3d and the B cell coreceptor (CD19/CD21) complex in the initiation of inflammatory arthritis. J Immunol 175(8):5379. https://doi.org/10.4049/jimmunol.175.8.5379

Deng L, Kim JR, Chang TZ, Zhang H, Mohan T, Champion JA, Wang BZ (2017) Protein nanoparticle vaccine based on flagellin carrier fused to influenza conserved epitopes confers full protection against influenza A virus challenge. Virology 509:82–89. https://doi.org/10.1016/j.virol.2017.06.001

Deng L, Mohan T, Chang TZ, Gonzalez GX, Wang Y, Kwon Y-M, Kang S-M, Compans RW, Champion JA, Wang B-Z (2018a) Double-layered protein nanoparticles induce broad protection against divergent influenza A viruses. Nat Commun 9(1):359. https://doi.org/10.1038/s41467-017-02725-4

Deng L, Chang TZ, Wang Y, Li S, Wang S, Matsuyama S, Yu G, Compans RW, Li J-D, Prausnitz MR, Champion JA, Wang B-Z (2018b) Heterosubtypic influenza protection elicited by double-layered polypeptide nanoparticles in mice. Proc Nat Acad Sci 115(33):E7758. https://doi.org/10.1073/pnas.1805713115

Doan CD, Ghosh S (2019) Formation and stability of pea proteins nanoparticles using ethanol-induced desolvation. Nanomaterials 9(7). https://doi.org/10.3390/nano9070949

Dong X, Liang J, Yang A, Qian Z, Kong D, Lv F (2019) A visible codelivery nanovaccine of antigen and adjuvant with self-carrier for cancer immunotherapy. ACS Appl Mater Interfaces 11(5):4876–4888. https://doi.org/10.1021/acsami.8b20364

Ekiert DC, Friesen RHE, Bhabha G, Kwaks T, Jongeneelen M, Yu WL, Ophorst C, Cox F, Korse H, Brandenburg B, Vogels R, Brakenhoff JPJ, Kompier R, Koldijk MH, Cornelissen L, Poon LLM, Peiris M, Koudstaal W, Wilson IA, Goudsmit J (2011) A highly conserved neutralizing epitope on group 2 influenza A viruses. Science 333(6044):843–850. https://doi.org/10.1126/science.1204839

El Bakkouri K, Descamps F, De Filette M, Smet A, Festjens E, Birkett A, Van Rooijen N, Verbeek S, Fiers W, Saelens X (2011) Universal vaccine based on ectodomain of matrix protein 2 of influenza A: Fc receptors and alveolar macrophages mediate protection. J Immunol 186(2):1022. https://doi.org/10.4049/jimmunol.0902147

Estrada LH, Chu S, Champion JA (2014) Protein nanoparticles for intracellular delivery of therapeutic enzymes. J Pharm Sci-Us 103(6):1863–1871. https://doi.org/10.1002/jps.23974

Ezzat K, Pernemalm M, Pålsson S, Roberts TC, Järver P, Dondalska A, Bestas B, Sobkowiak MJ, Levänen B, Sköld M, Thompson EA, Saher O, Kari OK, Lajunen T, Sverremark Ekström E, Nilsson C, Ishchenko Y, Malm T, Wood MJA, Power UF, Masich S, Lindén A, Sandberg JK, Lehtiö J, Spetz A-L, El Andaloussi S (2019) The viral protein corona directs viral pathogenesis and amyloid aggregation. Nat Commun 10(1):2331. https://doi.org/10.1038/s41467-019-10192-2

Feng J, Zhang M, Mozdzanowska K, Zharikova D, Hoff H, Wunner W, Couch RB, Gerhard W (2006) Influenza A virus infection engenders a poor antibody response against the ectodomain of matrix protein 2. Virol J 3:102. https://doi.org/10.1186/1743-422x-3-102

Fleischer CC, Payne CK (2014) Secondary structure of corona proteins determines the cell surface receptors used by nanoparticles. J Phys Chem B 118(49):14017–14026. https://doi.org/10.1021/jp502624n

Gao R, Cao B, Hu Y, Feng Z, Wang D, Hu W, Chen J, Jie Z, Qiu H, Xu K, Xu X, Lu H, Zhu W, Gao Z, Xiang N, Shen Y, He Z, Gu Y, Zhang Z, Yang Y, Zhao X, Zhou L, Li X, Zou S, Zhang Y, Li X, Yang L, Guo J, Dong J, Li Q, Dong L, Zhu Y, Bai T, Wang S, Hao P, Yang W, Zhang Y, Han J, Yu H, Li D, Gao GF, Wu G, Wang Y, Yuan Z, Shu Y (2013) Human infection with a novel avian-origin influenza A (H7N9) virus. N Engl J Med 368(20):1888–1897. https://doi.org/10.1056/NEJMoa1304459

Gregory AE, Titball R, Williamson D (2013) Vaccine delivery using nanoparticles. Front Cell Infect Mi 3. Unsp 13. https://doi.org/10.3389/fcimb.2013.00013

Gross O, Thomas CJ, Guarda G, Tschopp J (2011) The inflammasome: an integrated view. Immunol Rev 243(1):136–151. https://doi.org/10.1111/j.1600-065X.2011.01046.x

Hedstrom L (2002) Serine protease mechanism and specificity. Chem Rev 102(12):4501–4524. https://doi.org/10.1021/cr000033x

Herrera Estrada L, Wu H, Ling K, Zhang G, Sumagin R, Parkos CA, Jones RM, Champion JA, Neish AS (2017) Bioengineering bacterially derived immunomodulants: a therapeutic approach to inflammatory bowel disease. ACS Nano 11(10):9650–9662. https://doi.org/10.1021/acsnano.7b03239

Hornung V, Bauernfeind F, Halle A, Samstad EO, Kono H, Rock KL, Fitzgerald KA, Latz E (2008) Silica crystals and aluminum salts activate the NALP3 inflammasome through phagosomal destabilization. Nat Immunol 9(8):847–856. https://doi.org/10.1038/ni.1631

Huang RX, Carney RP, Stellacci F, Lau BLT (2013) Protein-nanoparticle interactions: the effects of surface compositional and structural heterogeneity are scale dependent. Nanoscale 5 (15):6928–6935. https://doi.org/10.1039/C3nr02117c

Hughey PG, Roberts PC, Holsinger LJ, Zebedee SL, Lamb RA, Compans RW (1995) Effects of antibody to the influenza A virus M2 protein on M2 surface expression and virus assembly. Virology 212(2):411–421. https://doi.org/10.1006/viro.1995.1498

Jahanshahi MB (2008) Z, Protein nanoparticle: a unique system as drug delivery vehicles. Afr J Biotechnol 7(25):4926–4934

Jegerlehner A, Storni T, Lipowsky G, Schmid M, Pumpens P, Bachmann MF (2002) Regulation of IgG antibody responses by epitope density and CD21-mediated costimulation. Eur J Immunol 32 (11):3305–3314. https://doi.org/10.1002/1521-4141(200211)32:11%3c3305:Aid-immu3305%3e3.0.Co;2-j

Jegerlehner A, Schmitz N, Storni T, Bachmann MF (2004) Influenza A vaccine based on the extracellular domain of M2: weak protection mediated via antibody-dependent NK cell activity. J Immunol 172(9):5598–5605. https://doi.org/10.4049/jimmunol.172.9.5598

Kallewaard NL, Corti D, Collins PJ, Neu U, McAuliffe JM, Benjamin E, Wachter-Rosati L, Palmer-Hill FJ, Yuan AQ, Walker PA, Vorlaender MK, Bianchi S, Guarino B, De Marco A, Vanzetta F, Agatic G, Foglierini M, Pinna D, Fernandez-Rodriguez B, Fruehwirth A, Silacci C, Ogrodowicz RW, Martin SR, Sallusto F, Suzich JA, Lanzavecchia A, Zhu Q, Gamblin SJ, Skehel JJ (2016) Structure and function analysis of an antibody recognizing all influenza A subtypes. Cell 166(3):596–608. https://doi.org/10.1016/j.cell.2016.05.073

Kanekiyo M, Wei CJ, Yassine HM, McTamney PM, Boyington JC, Whittle JRR, Rao SS, Kong WP, Wang LS, Nabel GJ (2013) Self-assembling influenza nanoparticle vaccines elicit broadly neutralizing H1N1 antibodies. Nature 499(7456):102–106. https://doi.org/10.1038/nature12202

Karch CP, Matyas GR, Burkhard P, Beck Z (2018) Self-assembling protein nanoparticles: implications for HIV-1 vaccine development. Nanomedicine 13(17):2121–2125. https://doi.org/10.2217/nnm-2018-0222

Keuth J, Nitschke Y, Mulac D, Riehemann K, Rutsch F, Langer K (2020) Reversion of arterial calcification by elastin-targeted DTPA-HSA nanoparticles. Eur J Pharm Biopharm 150:108–119. https://doi.org/10.1016/j.ejpb.2020.03.007

Kim Y-M, Pan JY-J, Korbel GA, Peperzak V, Boes M, Ploegh HL (2006) Monovalent ligation of the B cell receptor induces receptor activation but fails to promote antigen presentation. Proc Nat Acad Sci USA 103(9):3327. https://doi.org/10.1073/pnas.0511315103

Kim JR, Holbrook BC, Hayward SL, Blevins LK, Jorgensen MJ, Kock ND, De Paris K, D'Agostino RB, Aycock ST, Mizel SB, Parks GD, Alexander-Miller MA (2015) Inclusion of flagellin during vaccination against influenza enhances recall responses in nonhuman primate neonates. J Virol 89(14):7291–7303

Kirchdoerfer RN, Wang N, Pallesen J, Wrapp D, Turner HL, Cottrell CA, Corbett KS, Graham BS, McLellan JS, Ward AB (2018) Stabilized coronavirus spikes are resistant to conformational changes induced by receptor recognition or proteolysis. Sci Rep 8(1):15701. https://doi.org/10.1038/s41598-018-34171-7

Kisselev AF, Songyang Z, Goldberg AL (2000) Why does threonine, and not serine, function as the active site nucleophile in proteasomes? J Biol Chem 275(20):14831–14837. https://doi.org/10.1074/jbc.275.20.14831

Krammer F, Palese P (2013) Influenza virus hemagglutinin stalk-based antibodies and vaccines. Curr Opin Virol 3(5):521–530. https://doi.org/10.1016/j.coviro.2013.07.007

Kreuter J (1991) Nanoparticle-based dmg delivery systems. J Controlled Release 16(1):169–176. https://doi.org/10.1016/0168-3659(91)90040-K

Kumar S, Anselmo AC, Banerjee A, Zakrewsky M, Mitragotri S (2015) Shape and size-dependent immune response to antigen-carrying nanoparticles. J Control Release 220(Pt A):141–148. https://doi.org/10.1016/j.jconrel.2015.09.069

Kumar S, Sunagar R, Gosselin E (2019) Bacterial protein toll-like-receptor agonists: a novel perspective on vaccine adjuvants. Front Immunol 10:8. https://doi.org/10.3389/fimmu.2019.01144

Lai S, Qin Y, Cowling BJ, Ren X, Wardrop NA, Gilbert M, Tsang TK, Wu P, Feng L, Jiang H, Peng Z, Zheng J, Liao Q, Li S, Horby PW, Farrar JJ, Gao GF, Tatem AJ, Yu H (2016) Global epidemiology of avian influenza A H5N1 virus infection in humans, 1997-2015: a systematic review of individual case data. The Lancet. Infect Dis 16(7):e108–e118. https://doi.org/10.1016/S1473-3099(16)00153-5

Lamb RA, Zebedee SL, Richardson CD (1985) Influenza virus M2 protein is an integral membrane protein expressed on the infected-cell surface. Cell 40(3):627–633. https://doi.org/10.1016/0092-8674(85)90211-9

Langer K, Balthasar S, Vogel V, Dinauer N, von Briesen H, Schubert D (2003) Optimization of the preparation process for human serum albumin (HSA) nanoparticles. Int J Pharm 257(1–2):169–180. https://doi.org/10.1016/S0378-5173(03)00134-0

Li HF, Willingham SB, Ting JPY, Re F (2008) Cutting edge: inflammasome activation by alum and alum's adjuvant effect are mediated by NLRP3. J Immunol 181(1):17–21

Li W, Joshi MD, Singhania S, Ramsey KH, Murthy AK (2014) Peptide vaccine: progress and challenges. Vaccines 2(3):515–536. https://doi.org/10.3390/vaccines2030515

Lindman S, Lynch I, Thulin E, Nilsson H, Dawson KA, Linse S (2007) Systematic investigation of the thermodynamics of HSA adsorption to N-iso-propylacrylamide/N-tert-butylacrylamide copolymer nanoparticles. Effects of particle size and hydrophobicity. Nano Lett 7(4):914–920. https://doi.org/10.1021/nl062743

Liu L, Ma P, Wang H, Zhang C, Sun H, Wang C, Song C, Leng X, Kong D, Ma G (2016) Immune responses to vaccines delivered by encapsulation into and/or adsorption onto cationic lipid-PLGA hybrid nanoparticles. J Control Release 225:230–239. https://doi.org/10.1016/j.jconrel.2016.01.050

Luebbert CC, Clarke TM, Pointet R, Frahm GE, Tam S, Lorbetskie B, Sauve S, Johnston MJW (2017) Nanoparticle size and production efficiency are affected by the presence of fatty acids during albumin nanoparticle fabrication. Plos One 12(12). https://doi.org/10.1371/journal.pone.0189814

Mallajosyula VVA, Citron M, Ferrara F, Lu X, Callahan C, Heidecker GJ, Sarma SP, Flynn JA, Temperton NJ, Liang X, Varadarajan R (2014) Influenza hemagglutinin stem-fragment immunogen elicits broadly neutralizing antibodies and confers heterologous protection. Proc Nat Acad Sci 111(25):E2514. https://doi.org/10.1073/pnas.1402766111

Manolova V, Flace A, Bauer M, Schwarz K, Saudan P, Bachmann MF (2008) Nanoparticles target distinct dendritic cell populations according to their size. Eur J Immunol 38(5):1404–1413. https://doi.org/10.1002/eji.200737984

Markovsky E, Koroukhov N, Golomb G (2007) Additive-free albumin nanoparticles of alendronate for attenuating inflammation through monocyte inhibition. Nanomedicine 2 (4):545–553. https://doi.org/10.2217/17435889.2.4.545

Marty JJ, Oppenheim RC, Speiser P (1978) Nanoparticles—a new colloidal drug delivery system. Pharm Acta Helv 53(1):17–23

Mohammad-Beigi H, Shojaosadati SA, Morshedi D, Mirzazadeh N, Arpanaei A (2016) The effects of organic solvents on the physicochemical properties of human serum albumin nanoparticles. Iran J Biotechnol 14(1):45–50. https://doi.org/10.15171/ijb.1168

Mottram PL, Leong D, Crimeen-Irwin B, Gloster S, Xiang SD, Meanger J, Ghildyal R, Vardaxis N, Plebanski M (2007) Type 1 and 2 immunity following vaccination is influenced by nanoparticle size: formulation of a model vaccine for respiratory syncytial virus. Mol Pharmaceut 4(1):73–84. https://doi.org/10.1021/Mp060096p

Mozdzanowska K, Furchner M, Zharikova D, Feng J, Gerhard W (2005) Roles of CD4 + T-cell-independent and -dependent antibody responses in the control of influenza virus infection: evidence for noncognate CD4 + T-cell activities that enhance the therapeutic activity of antiviral antibodies. J Virol 79(10):5943–5951. https://doi.org/10.1128/jvi.79.10.5943-5951.2005

Munoz-Wolf N, Lavelle EC (2018) A guide to IL-1 family cytokines in adjuvanticity. FEBS J 285 (13):2377–2401. https://doi.org/10.1111/febs.14467

Murphy K, Travers P, Walport M, Janeway C (2012) Janeway's immunobiology. Garland Science, New York

Neumann S, Burkert K, Kemp R, Rades T, Dunbar PR, Hook S (2014) Activation of the NLRP3 inflammasome is not a feature of all particulate vaccine adjuvants. Immunol Cell Biol 92 (6):535–542. https://doi.org/10.1038/icb.2014.21

Oh JZ, Ravindran R, Chassaing B, Carvalho FA, Maddur MS, Bower M, Hakimpour P, Gill KP, Nakaya HI, Yarovinsky F, Sartor RB, Gewirtz AT, Pulendran B (2014) TLR5-mediated sensing of gut microbiota is necessary for antibody responses to seasonal influenza vaccination. Immunity 41(3):478–492

Reddy ST, van der Vlies AJ, Simeoni E, Angeli V, Randolph GJ, O'Neill CP, Lee LK, Swartz MA, Hubbell JA (2007) Exploiting lymphatic transport and complement activation in nanoparticle vaccines. Nat Biotechnol 25(10):1159–1164. https://doi.org/10.1038/Nbt1332

Roh YG, Shin SW, Kim SY, Kim S, Lim YT, Oh BK, Um SH (2019) Protein nanoparticle fabrication for optimized reticuloendothelial system evasion and tumor accumulation. Langmuir 35(11):3992–3998. https://doi.org/10.1021/acs.langmuir.8b03776

Roldão A, Mellado MC, Castilho LR, Carrondo MJ, Alves PM (2010) Virus-like particles in vaccine development. Expert Rev Vaccines 9(10):1149–1176. https://doi.org/10.1586/erv.10.115

Russell CJ, Hu M, Okda FA (2018) Influenza hemagglutinin protein stability, activation, and pandemic risk. Trends Microbiol 26(10):841–853. https://doi.org/10.1016/j.tim.2018.03.005

Schellekens H, Jiskoot W (2013) Pharm Biotech (Springer)

Sharma P, Hu-Lieskovan S, Wargo JA, Ribas A (2017) Primary, adaptive, and acquired resistance to cancer immunotherapy. Cell 168(4):707–723. https://doi.org/10.1016/j.cell.2017.01.017

Sharp FA, Ruane D, Claass B, Creagh E, Harris J, Malyala P, Singh M, O'Hagan DT, Petrilli V, Tschopp J, O'Neill LAJ, Lavelle EC (2009) Uptake of particulate vaccine adjuvants by dendritic cells activates the NALP3 inflammasome. Proc Natl Acad Sci USA 106(3):870–875. https://doi.org/10.1073/pnas.0804897106

Siegel S, Wagner A, Kabelitz D, Marget M, Coggin J Jr, Barsoum A, Rohrer J, Schmitz N, Zeis M (2003) Induction of cytotoxic T-cell responses against the oncofetal antigen-immature laminin receptor for the treatment of hematologic malignancies. Blood 102(13):4416–4423. https://doi.org/10.1182/blood-2003-01-0198

Singh S (2007) Vaccine adjuvants and delivery systems. Wiley-Interscience, Hoboken, N.J.

Snapper CM (2018) Distinct immunologic properties of soluble versus particulate antigens. Front Immunol 9:598–598. https://doi.org/10.3389/fimmu.2018.00598

Stano A, Nembrini C, Swartz MA, Hubbell JA, Simeoni E (2012) Nanoparticle size influences the magnitude and quality of mucosal immune responses after intranasal immunization. Vaccine 30(52):7541–7546. https://doi.org/10.1016/j.vaccine.2012.10.050

Storp B, Engel A, Boeker A, Ploeger M, Langer K (2012) Albumin nanoparticles with predictable size by desolvation procedure. J Microencapsul 29(2):138–146. https://doi.org/10.3109/02652048.2011.635218

Subara D, Jaswir I, Alkhatib MFR, Noorbatcha IAN (2017) Process optimization for the production of fish gelatin nanoparticles. Int Food Res J 24:S501–S507

Tisoncik JR, Korth MJ, Simmons CP, Farrar J, Martin TR, Katze MG (2012) Into the eye of the cytokine storm. Microbiol Mol Biol Rev MMBR 76(1):16–32. https://doi.org/10.1128/MMBR.05015-11

Tompkins SM, Zhao Z-S, Lo C-Y, Misplon JA, Liu T, Ye Z, Hogan RJ, Wu Z, Benton KA, Tumpey TM, Epstein SL (2007) Matrix protein 2 vaccination and protection against influenza viruses, including subtype H5N1. Emerg Infect Dis 13(3):426–435. https://doi.org/10.3201/eid1303.061125

Tran KK, Shen H (2009) The role of phagosomal pH on the size-dependent efficiency of cross-presentation by dendritic cells. Biomaterials 30(7):1356–1362. https://doi.org/10.1016/j.biomaterials.2008.11.034

Treanor JJ, Tierney EL, Zebedee SL, Lamb RA, Murphy BR (1990) Passively transferred monoclonal antibody to the M2 protein inhibits influenza A virus replication in mice. J Virol 64(3):1375–1377

Tsoras AN, Champion JA (2018) Cross-linked peptide nanoclusters for delivery of oncofetal antigen as a cancer vaccine. Bioconjug Chem 29(3):776–785. https://doi.org/10.1021/acs.bioconjchem.8b00079

Tsoras AN, Wong KM, Paravastu AK, Champion JA (2020) Rational design of antigen incorporation into subunit vaccine biomaterials can enhance antigen-specific immune responses. Front Immunol. https://doi.org/10.3389/fimmu.2020.01547

Vaine CA, Patel MK, Zhu JT, Lee E, Finberg RW, Hayward RC, Kurt-Jones EA (2013) Tuning innate immune activation by surface texturing of polymer microparticles: the role of shape in inflammasome activation. J Immunol 190(7):3525–3532. https://doi.org/10.4049/jimmunol.1200492

Wang W, Singh SK, Li N, Toler MR, King KR, Nema S (2012) Immunogenicity of protein aggregates–concerns and realities. Int J Pharm 431(1–2):1–11. https://doi.org/10.1016/j.ijpharm.2012.04.040

Wang L, Hess A, Chang TZ, Wang YC, Champion JA, Compans RW, Wang BZ (2014) Nanoclusters self-assembled from conformation-stabilized influenza M2e as broadly cross-protective influenza vaccines. Nanomed-Nanotechnol 10(2):473–482. https://doi.org/10.1016/j.nano.2013.08.005

Wang L, Chang TZ, He Y, Kim JR, Wang S, Mohan T, Berman Z, Tompkins SM, Tripp RA, Compans RW, Champion JA, Wang B-Z (2017) Coated protein nanoclusters from influenza H7N9 HA are highly immunogenic and induce robust protective immunity. Nanomed Nanotechnol Biol Med 13(1):253–262. https://doi.org/10.1016/j.nano.2016.09.001

Weber C, Coester C, Kreuter J, Langer K (2000) Desolvation process and surface characterisation of protein nanoparticles. Int J Pharm 194(1):91–102. https://doi.org/10.1016/S0378-5173(99)00370-1

Willett TA, Meyer AL, Brown EL, Huber BT (2004) An effective second-generation outer surface protein A-derived Lyme vaccine that eliminates a potentially autoreactive T cell epitope. Proc Natl Acad Sci USA 101(5):1303–1308. https://doi.org/10.1073/pnas.0305680101

Wu P, Peng Z, Fang VJ, Feng L, Tsang TK, Jiang H, Lau EHY, Yang J, Zheng J, Qin Y, Li Z, Leung GM, Yu H, Cowling BJ (2016) Human infection with influenza A(H7N9) virus during 3 major epidemic waves, China, 2013–2015. Emerg Infect Dis 22(6):964–972. https://doi.org/10.3201/eid2206.151752

Xiang SD, Scholzen A, Minigo G, David C, Apostolopoulos V, Mottram PL, Plebanski M (2006) Pathogen recognition and development of particulate vaccines: does size matter? Methods 40(1):1–9. https://doi.org/10.1016/j.ymeth.2006.05.016

Xiang SD, Wilson K, Day S, Fuchsberger M, Plebanski M (2013) Methods of effective conjugation of antigens to nanoparticles as non-inflammatory vaccine carriers. Methods 60(3):232–241. https://doi.org/10.1016/j.ymeth.2013.03.036

Zhang Y, Xu C, Zhang H, Liu GD, Xue C, Cao Y (2019) Targeting hemagglutinin: approaches for broad protection against the influenza A virus. Viruses 11(5):405. https://doi.org/10.3390/v11050405

Zhao L, Seth A, Wibowo N, Zhao C-X, Mitter N, Yu C, Middelberg APJ (2014) Nanoparticle vaccines. Vaccine 32(3):327–337. https://doi.org/10.1016/j.vaccine.2013.11.069

Zheng M, Luo J, Chen Z (2014) Development of universal influenza vaccines based on influenza virus M and NP genes. Infection 42(2):251–262. https://doi.org/10.1007/s15010-013-0546-4

Zhu W, Pewin W, Wang C, Luo Y, Gonzalez GX, Mohan T, Prausnitz MR, Wang BZ (2017) A boosting skin vaccination with dissolving microneedle patch encapsulating M2e vaccine broadens the protective efficacy of conventional influenza vaccines. J Control Release 261:1–9. https://doi.org/10.1016/j.jconrel.2017.06.017

Zolnik BS, Gonzalez-Fernandez A, Sadrieh N, Dobrovolskaia MA (2010) Minireview: nanoparticles and the immune system. Endocrinology 151(2):458–465. https://doi.org/10.1210/En.2009-1082

The manufacturer's authorised representative in the EU is Springer Nature Customer Service Centre GmbH, Europaplatz 3, 69115 Heidelberg, Germany. If you have any concerns regarding our products, please contact ProductSafety@springernature.com

Printed and bound by CPI Group (UK) Ltd, Croydon, CR0 4YY
12/08/2026
02194540-0001